DOING MATHEMATICS

An Introduction to Proofs and Problem Solving

STEVEN GALOVICH

Carleton College

SAUNDERS COLLEGE PUBLISHING
Harcourt Brace Jovanovich College Publishers

*Fort Worth Philadelphia San Diego New York Orlando Austin San Antonio
Toronto Montreal London Sydney Tokyo*

Galovich: Doing Mathematics: An Introduction to Proofs and Problem Solving

ISBN 0-03-092800-1

345 066 987654321

*To my mother,
Helen Galovich,
and the memory
of my father,
Steve Galovich.*

PREFACE

This book is intended to introduce a budding mathematics student to the process of doing mathematics. Accordingly, it can be used in several ways: as a text for a short course on proofs and mathematical reasoning; as a supplemental text in a sophomore-level linear algebra course or a junior-level algebra or analysis course; or as independent reading for students at various levels.

As we all know, any first- or second-year college student who is serious about mathematics has been studying the subject for at least twelve years. Thus it might seem strange, or even pretentious, to claim that this book can provide such a student with an introduction to the mathematical process. Have not students already been introduced to mathematics? And if not, what were they doing in all those math classes in grade school and high school?

Most students who have taken a college calculus course intended for those destined to major in mathematics, science, or engineering (among other majors) will probably agree that this calculus course surpasses previous mathematics courses in sophistication and difficulty. Most teachers of these courses will also agree. In calculus and in subsequent mathematics courses, students are expected to perform beyond the levels they attained in high school in the following ways:

(1) They are expected to understand the underlying concepts of the subject.
(2) They should be able to apply these concepts to solving problems.
(3) They are expected to understand and perhaps to reproduce proofs of theorems.
(4) They should comprehend the connections among the important concepts and techniques of the subject.

For the most part, these expectations do not exist in high school mathematics courses. Thus college math courses often seem strange and bewildering to many students. The mathematics taught in college, especially in courses beyond calculus, seems to many students to be an entirely different kind of animal than the mathematics taught in high school, almost a completely new subject.

This book is intended to serve as an introduction to this "new" mathematics. In a sense, our main concern is with items (1) through (4) listed in the previous paragraph. Our hope is that by discussing proof techniques, problem solving methods, and the understanding of mathematical ideas, we can help the student successfully step into higher-level college mathematics courses.

The first unit of the book deals with logic and sets, two of the fundamental building blocks of mathematics. We study both propositional and predicate calculus, and in between, introduce informally the concept of a set.

In Unit Two, we consider methods of proving theorems. The principal proof techniques presented are direct proof, proof by contradiction, proof by contraposition, mathematical induction, and case analysis. Along the way, we discuss the general notion of an axiomatic system and give an axiomatic presentation of the real numbers. The unit closes with a return to basic set theory. Set operations are discussed in Section 10. The important concepts of relations and equivalence relations are presented in Sections 11 and 13 respectively. Section 12, which can be read immediately after Section 10, describes fundamental properties of functions.

The third unit looks at problem solving—methods that one can use to attack a problem. Our approach is based on the work of the mathematician George Polya. The techniques articulated by Polya are intended to be guidelines or rules-of-thumb that are helpful in attacking a problem, but do not in every case necessarily guarantee a solution. Nonetheless, Polya's "rules" are extremely valuable, not only for solving problems but also for understanding mathematical ideas.

In Unit Four, we consider some aspects of an individual's behavior and attitude that affect the way he or she actually learns and does mathematics. Following a path blazed by Alan Schoenfeld, we look at the beliefs that many people bring to the study of mathematics and to the way people behave when they work on mathematical problems. Unit Four is entirely nontechnical and can be read at any time. In fact, all students are encouraged to read Unit Four early and often. Finally, the book closes with a collection of problems. Our hope is that the spirit of doing mathematics is conveyed in this section. The problems are intended to be challenging, intriguing, and just plain fun.

Throughout this book, the symbol ∎ marks the end of a proof.

This book is a spinoff of the text *Introduction to Mathematical Structures* also published by Harcourt Brace Jovanovich. In fact this book consists of large parts of Chapters 1 and 2 of that text with some changes and additions.

As was the case with *Introduction to Mathematical Structures*, there are many intellectual and personal debts that I owe. The material presented in Units 3 and 4 relies heavily on the writings of George Polya and Alan Schoenfeld. Any reader of this book will enjoy Polya's *How to Solve It* and his other works cited in the bibliography, and at least the first half of Schoenfeld's *Mathematical Problem Solving*. Closer to home, I want to thank my Carleton colleagues, especially David Appleyard and Mark Krusemeyer, for their comments on *Introduction to Mathematical Struc-*

tures. Professor Krusemeyer has volunteered many useful comments that have found their way into this text, and as a problem poser *par excellence*, has suggested many interesting problems that are scattered throughout this book. I also wish to thank Professor William Cherowitzo of the University of Colorado at Denver for his many valuable suggestions.

Finally, I want to thank several people who are responsible for the production of the manuscript. Mike Tie and Peter Grauff produced the pretty figures and managed to have them land in the right spots in the text. Mike also performed a number of amazing technical acrobatics and enlisted a cadre of eager Carleton students who contributed to the creation of the figures. Included in this army of artists were Nick Coult, Loren Frank, Jeff Graves, John Littell, Chris Lochstet, Jack Nutting, Darcy Paquet, Karen Swanberg, and Loren Woo. Most of all, I wish to thank Barbara Jenkins for her expert typing and her wonderfully cheerful spirit through what turned out to be an extended production period. Thanks again, Barbara.

Lastly, I thank my wife, Elizabeth Culver, for her enthusiastic support for this project.

CONTENTS

UNIT 1
THINKING LOGICALLY   1

1. Propositional Calculus 2
2. Sets 13
3. Predicates and Quantifiers 27

UNIT 2
PROVING THEOREMS   33

4. Direct Proof 38
5. Indirect Proof 42
6. An Example: The Real Numbers 48
7. Mathematical Induction 61
8. Case Analysis 71
9. Quantification and Counterexamples . . . 77
10. Sets and Set Operations 81
11. Relations 98
12. Functions 108
13. Equivalence Relations 126

UNIT 3
SOLVING PROBLEMS   139

14. How to Solve It 139
15. Understanding the Problem 143

16. Attacking the Problem 155
17. Looking Back 166

UNIT 4

DOING MATHEMATICS 169

18. Controlling Your Thinking 169
19. Attitudes and Beliefs 172
20. Problems 176

Hints and Solutions to Selected Exercises . 185
References 194
Index 195

Unit 1
THINKING
LOGICALLY

Logic or symbolic logic is the study of valid reasoning processes. As such, logic provides the basic rules by which reasoning in mathematics is carried out. As Davis and Hersh point out in *The Mathematical Experience*, logic is the "nuts and bolts" of mathematics.

In this unit we present a very brief, utilitarian summary of logic. We concentrate first on the so-called propositional calculus, that part of logic that focuses on the analysis of statements that are either true or false. After a quick introduction to the idea of set theory, we venture into the domain of predicate calculus, which is concerned with "indefinite" statements (i.e., those that are not necessarily true or false) that describe properties of objects and relationships among objects. We emphasize that our purpose is not to study logic for its own sake but as a tool to carry out mathematical investigations.

A *proposition* is a sentence that is either true or false but is not both true and false. For example,

Barbara Jones has gray hair,

John Smith is at least six feet tall,

are examples of propositions. To each proposition, we assign a *truth value*, true or false, depending on whether the proposition is true or false. (As an aside, notice that not every sentence is a proposition. For example, here is a sentence to which a truth value cannot be assigned:

This statement is false.

A moment's thought reveals that if the statement is true, then it is false, and if the statement is false, then it is true. Thus, no truth value can be assigned to the statement; hence, the statement is not a proposition.)

We now turn to the *calculus* of propositions; that is, how new propositions can be formed from existing propositions. We list five ways of creating new propositions from old ones: conjunction, disjunction, negation, conditional, and biconditional. These fancy words refer to the creation of new propositions through the use of the words "and," "or," "not," "if-then," and "if and only if," respectively. These items are called *logical connectives*. In what follows, uppercase single letters will be used to denote propositions.

Definition 1 *Logical Connectives*

Let P and Q be propositions. The five logical connectives are defined as follows:
  i. The *conjunction* of P and Q is the statement "P and Q."
  ii. The *disjunction* of P and Q is the statement "P or Q."
  iii. The *negation* of P is the statement "not-P," which is often phrased as "it is not the case that P."
  iv. The *conditional* formed from the pair P and Q is the statement "if P then Q."
  v. The *biconditional* formed from the pair P and Q is the statement "P if and only if Q."

---

1. The *conjunction* of P and Q is the sentence "P and Q." A shorthand symbolic way of writing "P and Q" is $P \land Q$. For example, if P and Q are respectively the sentences given above, then "P and Q" is the sentence

Barbara Jones has gray hair and John Smith is at least six feet tall.

The conjunction of P and Q is defined to be true if P and Q are both true and is defined to be false in all other cases. A convenient way of conveying this information is via a device called a *truth table* The truth table for "P and Q" expresses the truth or falsity of "P and Q" in terms of the truth values of the individual sentences P and Q. Table 1 is the truth table for "P and Q."

P	Q	P and Q
T	T	T
T	F	F
F	T	F
F	F	F

Table 1

Reading horizontally we see that if P is true (T) and Q is true, then "P and Q" is true. If P is true and Q is false (F), then "P and Q" is false, etc.

As a quick exercise, let us make a truth table for the sentence "Q and P."

P	Q	Q and P
T	T	T
T	F	F
F	T	F
F	F	F

Table 2

Observe that the propositions "P and Q" and "Q and P" have the same truth values for each pair of possible truth values for Q and P.

2. The *disjunction* of P and Q is the sentence "P or Q"; "P or Q" is sometimes written $P \lor Q$. The disjunction of P and Q is defined to be true if at least one of P and Q is true and it is defined to be false if both P and Q are false. The truth table for "P or Q" is

P	Q	P or Q
T	T	T
T	F	T
F	T	T
F	F	F

Table 3

Observe that "*Q* or *P*" has the same truth values as "*P* or *Q*" for each pair of truth values of the propositions *P* and *Q*.

The disjunction "*P* or *Q*" is also known as the "inclusive-or." This is in contrast to the "exclusive-or" (xor) which in everyday discourse is often used in place of "or." The truth value for the exclusive-or is determined as follows: "*P* xor *Q*" is true when exactly one of "*P* and *Q*" is true. ("At the corner I will turn right or I will turn left" is an example of an exclusive-or.) In mathematics, the inclusive-or is used almost exclusively.

3. For a given proposition *P*, the *negation* of *P*, written "not-*P*," is defined to be true when *P* is false and is defined to be false when *P* is true. The truth table for "not-*P*" is

Table 4

The proposition "not-*P*" is expressed symbolically in various ways including $\neg P$, $\sim P$, and $-P$.

For each pair of propositions *P* and *Q*, no matter how complex, the propositions "*P* or *Q*" and "*P* and *Q*" are easy to phrase grammatically. Sometimes, however, it is difficult to give an accurate and literate rendition of "not-*P*". A cautious but legitimate way to state "not-*P*" is, "it is not the case that *P*." Thus, if *P* is the proposition "John is a boy," then "not-*P*" can be read as:

It is not the case that John is a boy,

which can in turn be translated into:

John is not a boy.

As another example, suppose *P* is the sentence:

John is a boy or Jane is a girl.

Then not-*P* reads:

It is not the case that John is a boy or Jane is a girl,

which evidently can be accurately restated as:

John is not a boy and Jane is not a girl.

Generalizing from this observation, we might conjecture (i.e., guess) that for any pair of propositions *P* and *Q*, the proposition "not-(*P* or *Q*)" has the same "meaning" as the proposition "(not-*P* and not-*Q*)." More precisely, we conjecture that "not-(*P* or *Q*)" and "not-*P* and not-*Q*" have the same truth tables:

P	Q	P or Q	not-$(P$ or $Q)$	not-P	not-Q	not-P and not-Q
T	T	T	F	F	F	F
T	F	T	F	F	T	F
F	T	T	F	T	F	F
F	F	F	T	T	T	T

Table 5

Thus no matter what the truth values of P and Q, the truth values of "not-$(P$ or $Q)$" and "not-P and not-Q" coincide. This observation supports the interpretation of "not-$(P$ or $Q)$" that was made in the previous paragraph.

We have thus observed several examples in which two different statements (in this example "not-$(P$ or $Q)$" and "(not-P) and (not-Q)") have identical truth tables. This phenomenon can be formalized into a definition.

Definition 2 *Logical Equivalence*

Suppose A and B are propositions formed from a given collection of propositions P, Q, R, S,... using "and," "or," "not," and the conditional and biconditional to be defined below. The propositions A and B are *logically equivalent* if A and B have the same truth values for all possible truth values of the propositions P, Q, R, S,....

Table 5 demonstrates that "not-$(P$ or $Q)$" and "(not-P) and (not-Q)" are logically equivalent.

The following table shows the logical equivalence of "not-$(P$ and $Q)$" and "(not-P) or (not-Q)":

P	Q	P and Q	not-$(P$ and $Q)$	not-P	not-Q	not-P or not-Q
T	T	T	F	F	F	F
T	F	F	T	F	T	T
F	T	F	T	T	F	T
F	F	F	T	T	T	T

Table 6

This result tells us how to translate the negation of a conjunction. To negate the conjunction, P and Q, simple negate each component (i.e., negate P and negate Q) and change the word "and" to "or." For example, the negation of

John is a boy and Jane is a girl

is

John is not a boy or Jane is not a girl.

We saw earlier that the propositions "P and Q" and "Q and P" are logically equivalent and that "P or Q" and "Q or P" are logically equivalent. Here are some other examples:

   i. P is logically equivalent to "P and P"
  ii. P is logically equivalent to "P or P"
 iii. "$(P$ and $Q)$ and R" is logically equivalent to "P and $(Q$ and $R)$."

Now back to the logical connectives.

4. The next method of combining a pair of propositions to form a new proposition is via the conditional statement. Let P and Q be propositions. The *conditional statement* "If P then Q" is true when P and Q are both true or when P is false; in the only remaining case, which occurs when P is true and Q is false, "If P then Q" is false.

The truth table for "If P then Q" is

P	Q	If P then Q
T	T	T
T	F	F
F	T	T
F	F	T

Table 7

The conditional "If P then Q" is also called an *implication* and is often written "P implies Q." Symbolically the implication "If P then Q" is written $P \Rightarrow Q$. In an implication or conditional $P \Rightarrow Q$, we refer to the proposition P as the *hypothesis* of the implication and the statement Q as the *conclusion* of the implication. Some people call P the *antecedent* and Q the *consequent* of the conditional $P \Rightarrow Q$.

Notice that the implication $P \Rightarrow Q$ is defined to be true whenever P is false. This practice might seem puzzling, but there are good reasons for it. First, why should we consider $P \Rightarrow Q$ to be true when P is false and Q is true? Perhaps the most reasonable response is that any implication with a true conclusion should be regarded as true no matter what the hypothesis. Thus,

If I have 1,000,000 dollars, then I have 5 cents

is true since I do indeed have 5 cents even though I do not have 1,000,000 dollars. What about when P and Q are both false? One justification for defining $P \Rightarrow Q$ to be true in this case is that we want the implication $P \Rightarrow P$ to be true for any proposition P. Thus

If I have 1,000,000 dollars, then I have 1,000,000 dollars

is true even if I do not have 1,000,000 dollars. Thus to ensure that $P \Rightarrow P$ is true for all propositions P, we define $P \Rightarrow Q$ to be true when P and Q are both false.

Let "If P then Q" be a given implication. The *contrapositive* of "If P then Q" is defined to be the conditional "If not-Q then not-P." For example, the contrapositive of the implication

$$\text{If } x = 2, \text{ then } x^2 = 4$$

is the implication

$$\text{If } x^2 \neq 4, \text{ then } x \neq 2.$$

How does the contrapositive of a conditional compare logically with the conditional itself? To answer this question we form a truth table.

P	Q	$P \Rightarrow Q$	not-Q	not-P	$(\text{not-}Q) \Rightarrow (\text{not-}P)$
T	T	T	F	F	T
T	F	F	T	F	F
F	T	T	F	T	T
F	F	T	T	T	T

Table 8

As is evident in the table, for each pair of possible truth values of P and Q, the truth values of $P \Rightarrow Q$ and $(\text{not-}Q) \Rightarrow (\text{not-}P)$ coincide. Thus, any implication is logically equivalent to its contrapositive.

(This remark is more than just an idle curiosity. Suppose that we are trying to prove that $P \Rightarrow Q$ is true. This implication holds exactly when the contrapositive $(\text{not-}Q) \Rightarrow (\text{not-}P)$ holds. Moreover, in a given case, the latter implication might be easier to establish than the former. Thus, the logical equivalence of $P \Rightarrow Q$ and $(\text{not-}Q) \Rightarrow (\text{not-}P)$ actually provides us with a method of proving that $P \Rightarrow Q$ is true. As we shall see in Unit II, this proof technique will become one of the tools used to establish mathematical statements.)

Our next observation concerning implications leads to a question. Notice that "If P then Q" is true for three of the possible four pairs of truth values of P and Q. Similarly, the disjunction of two propositions is true for three of the four pairs of truth values of the component propositions. Thus, a connection between implications and disjunctions is suggested. Specifically, we ask: Is there a disjunction formed somehow from P and Q that is logically equivalent to the implication "If P then Q"? It is certainly not "P or Q." (Why not?) What then is it? Since "If P then Q" is false when P is true and Q is false, it is reasonable to consider a disjunction that is false when P is true and Q is false; "not-P or Q" is such a disjunction. As an exercise, verify that "If P then Q" and "not-P or Q" are logically equivalent.

A corollary of the last remark concerns the negation of "If P then Q." Since $P \Rightarrow Q$ is logically equivalent to not-$P \vee Q$, not-$(P \Rightarrow Q)$ is logically equivalent to not-(not-$P \vee Q$), which in turn is logically equivalent to (not-(not-P))$\wedge$(not-Q) by Table 5. Since not-(not-P) is logically equivalent to P, not-$(P \Rightarrow Q)$ is logically equivalent to $P \wedge (\text{not-}Q)$.

Let $P \Rightarrow Q$ be a given implication. The *converse* of $P \Rightarrow Q$ is the conditional $Q \Rightarrow P$. For instance, the converse of the implication

$$\text{If } x = 2, \text{ then } x^2 = 4$$

is the implication

$$\text{If } x^2 = 4, \text{ then } x = 2.$$

Are the implications "If P then Q" and its converse "If Q then P" logically equivalent? Let us form a truth table:

P	Q	$P \Rightarrow Q$	$Q \Rightarrow P$
T	T	T	T
T	F	F	T
F	T	T	F
F	F	T	T

Table 9

Because $P \Rightarrow Q$ and $Q \Rightarrow P$ have different truth values when P is true and Q is false, *these propositions are not logically equivalent.*

This observation actually opens the way for a wealth of mathematical questions. Whenever we prove a statement of the form "If P then Q," we can ask: Is the converse "If Q then P" also true? For certain pairs of statements, P and Q, the converse will indeed hold; in other cases it will not. For example, the statement "for a real number x, if $x + 4 = 8$, then $x = 4$" and its converse are both true. On the other hand, the following are true statements whose converses are false:

i. For a real number x, if $x = 2$, then $x^2 = 4$.
ii. For triangles T_1 and T_2 in the Euclidean plane, if T_1 and T_2 are congruent, then corresponding angles of T_1 and T_2 are equal.

Thus, the moral is that whenever we prove or even attempt to prove an implication $P \Rightarrow Q$, we should also consider its converse $Q \Rightarrow P$. Both $P \Rightarrow Q$ and $Q \Rightarrow P$ might be true, in which case P and Q provide two ways to express the same concept (see 5 below). For example, to say about a real number x that $x = 4$ is equivalent to saying that $x + 4 = 8$. If only one of the implications $P \Rightarrow Q$ and $Q \Rightarrow P$ is true, then we obtain a clearer understanding of the relationship between P and Q.

5. The last logical connective to be considered is the *biconditional*. By definition the biconditional of P and Q is the statement "(if P then Q) and (if Q then P)." In logical symbols the biconditional is $(P \Rightarrow Q) \wedge (Q \Rightarrow P)$.

Symbolically the biconditional of P and Q is written $P \Leftrightarrow Q$ and is read "P if and only if Q." The statement "P if and only if Q" is sometimes abbreviated "P iff Q."

Under what conditions is $P \Leftrightarrow Q$ true? By definition $P \Leftrightarrow Q$ is true precisely when the implication $P \Rightarrow Q$ and its converse are true. We consider two cases: (i) P is true, (ii) P is false.

(i) If P is true, then since $P \Rightarrow Q$ is true, Q must also be true. (ii) If P is false, then since $Q \Rightarrow P$ is true, Q must also be false. Thus, if the implications "if P then Q" and "if Q then P" are both true, then P and Q must have the same truth values; i.e., P and Q must be logically equivalent. On the other hand if P and Q are either both true or both false, then $P \Rightarrow Q$ and $Q \Rightarrow P$ are both true. Thus, to summarize our findings, the implications $P \Rightarrow Q$ and $Q \Rightarrow P$ are both true precisely when P and Q have the same truth values. In other words, $P \Leftrightarrow Q$ is true when P and Q are logically equivalent and false when P and Q are not logically equivalent. The truth table for the biconditional is

P	Q	$P \Leftrightarrow Q$
T	T	T
T	F	F
F	T	F
F	F	T

Table 10

Both conditionals and biconditionals are often phrased in other ways. Each of these different forms occurs frequently enough to warrant mentioning at this time.

(i) "P if Q" means "if Q then P"
(ii) "P only if Q" means "if P then Q"
(iii) "P is a necessary condition for Q" means "if Q then P"
(iv) "P is a sufficient condition for Q" means "if P then Q"
(v) "P is a necessary and sufficient condition for Q" means "P iff Q"

Let us now apply the rules concerning truth values of propositions to some specific cases.

Example 1 Write out a truth table for $(P \Rightarrow Q) \Rightarrow (Q \Rightarrow P)$. The table is

P	Q	$P \Rightarrow Q$	$Q \Rightarrow P$	$(P \Rightarrow Q) \Rightarrow (Q \Rightarrow P)$
T	T	T	T	T
T	F	F	T	T
F	T	T	F	F
F	F	T	T	T

Table 11

Notice that $(P \Rightarrow Q) \Rightarrow (Q \Rightarrow P)$ and $Q \Rightarrow P$ are logically equivalent.

Example 2   Show that $P \Leftrightarrow Q$ and "$(P$ and $Q)$ or (not-P and not-Q)" are logically equivalent.

To show the logical equivalence of the given propositions, we construct a truth table for each.

P	Q	$P \Leftrightarrow Q$	P and Q	(not-P) and (not-Q)	$(P$ and $Q)$ or (not-P and not-Q)
T	T	T	T	F	T
T	F	F	F	F	F
F	T	F	F	F	F
F	F	T	F	T	T

Table 12

Example 3   Construct a truth table for $[P \vee ((\text{not-}P) \wedge Q)] \vee [\text{not-}P \wedge \text{not-}Q]$.

P	Q	$P \vee ((\text{not-}P) \wedge Q)$	(not-P) $\wedge$ (not-Q)	$[P \vee ((\text{not-}P) \wedge Q)]$ $\vee[\text{not-}P \wedge \text{not-}Q]$
T	T	T	F	T
T	F	T	F	T
F	T	T	F	T
F	F	F	T	T

Table 13

Thus, the given proposition is true no matter what the truth values of P and Q.

Definition 3   *Tautology and Contradiction*

   Let A be a proposition formed from propositions $P, Q, R, \ldots$ using the logical connectives.
   (a) A is called a *tautology* if A is true for every assignment of truth values to $P, Q, R, \ldots$.
   (b) A is called a *contradiction* if A is false for every assignment of truth values to $P, Q, R, \ldots$.

For example, as Table 13 shows, $[P \vee ((\text{not-}P) \wedge Q)] \vee [\text{not-}P \wedge \text{not-}Q]$ is tautology. On the other hand, from Example 1 we see that $(P \Rightarrow Q) \Rightarrow (Q \Rightarrow P)$ is not a tautology. Notice that two propositions A and B (formed from propositions $P, Q, R, \ldots$) are logically equivalent if and only if the biconditional $A \Leftrightarrow B$ is a tautology. Finally, observe that A is a tautology if and only if not-A is a contradiction.

The examples given above provide some practice with truth tables. Our goal, however, is not to make us all truth-table wizards. Instead, we should acquire solid knowledge of tables 1, 3, 4, 7, and 10 (working problems such as the ones given above helps to develop this skill) and we should develop the ability to form the negation of a statement and the converse and contrapositive of an implication.

EXERCISES §1

1. Let P, Q, and R denote propositions. Construct truth tables for each of the following propositions.
   (a) $P \Rightarrow (P \Rightarrow Q)$
   (b) $(P \Rightarrow Q) \Rightarrow (P \wedge \text{not-}Q)$
   (c) $P \Rightarrow (\text{not-}(Q \text{ and } R))$
   (d) $(P \Rightarrow (Q \text{ and } R))$ or $(\text{not-}P \text{ and } Q)$
   (e) $(P \text{ and } Q) \Rightarrow [((Q \text{ and } (\text{not-}Q)) \Rightarrow (R \text{ and } Q))]$
   (f) $(P \Rightarrow Q) \Rightarrow (\text{not-}P \Rightarrow \text{not-}Q)$
2. Use truth tables to show that for any statement P, P is logically equivalent to each of the following: (i) P and P, (ii) P or P, (iii) not-$(\text{not-}P)$.
3. (a) Suppose P and $P \Rightarrow Q$ are true. Show Q is true.
   (b) Suppose P and Q are propositions for which "P and Q" is false and "P or Q" is true. What can be said about the truth values of P and Q?
   (c) If $P \Rightarrow Q$ is true for all propositions Q, then what can be said about the truth value of P?
   (d) If $P \Leftrightarrow Q$ is true, what is the truth value of $P \Leftrightarrow (\text{not-}Q)$?
4. Are $P \Rightarrow Q$ and $(\text{not-}P) \Rightarrow (\text{not-}Q)$ logically equivalent?
5. Prove by constructing a truth table that "not-$P \vee Q$" is logically equivalent to $P \Rightarrow Q$.
6. (a) Show that not-$(P \Rightarrow Q)$ is logically equivalent to "P and $(\text{not-}Q)$".
   (b) Show that $P \Rightarrow (Q \vee R)$ is logically equivalent to $(P \Rightarrow Q) \vee (P \Rightarrow R)$.
7. Let P, Q, and R be propositions. Show that the following pairs of statements are logically equivalent. (In every case your truth table will have eight rows.)
   (a) $(P \wedge Q) \wedge R$ and $P \wedge (Q \wedge R)$.
   (b) $(P \vee Q) \vee R$ and $P \vee (Q \vee R)$.
8. Show that each of the following is a tautology:
   (a) P or not-P.
   (b) If P then P.
9. Show that $(P \Leftrightarrow Q) \Leftrightarrow [(P \wedge Q) \vee ((\text{not-}P) \wedge (\text{not-}Q))]$ is a tautology.
10. Prove that $((P \Rightarrow Q) \wedge P) \Rightarrow Q$ is a tautology.
11. Show that $P \Rightarrow (Q \vee R)$ is logically equivalent to $(P \Rightarrow Q) \vee (P \Rightarrow R)$.
12. Let P and Q be propositions.

(a) Construct a truth table for the proposition $(P \lor Q) \Rightarrow (P \land \sim Q)$.

(b) Find a simpler proposition that is logically equivalent to $(P \lor Q) \Rightarrow (P \land \sim Q)$.

13. Is $P \Rightarrow (\text{not-}P)$ a tautology? Explain.

14. Let us define a new logical connective $\triangledown$, called *nor*, by the following truth table:

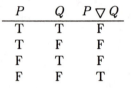

P	Q	$P \triangledown Q$
T	T	F
T	F	F
F	T	F
F	F	T

(a) Prove that not-P is logically equivalent to $P \triangledown P$.

(b) Find a proposition that contains only P, Q, parentheses, and the connective $\triangledown$ (some or all of which may appear more than once) and which is logically equivalent to $P \lor Q$.

(c) Do the same for $P \land Q$.

(d) Do the same for $P \Rightarrow Q$.

15. Complete each of the following sentences:

(a) The negation of the conjunction of two propositions is the .... .

(b) The negation of the disjunction of two propositions is the .... .

Section 2
SETS

As a discipline within mathematics, the theory of sets had its origins in the study of functions carried out during the latter half of the nineteenth century. Since then, set theory has developed into a full-blown area of specialization inside mathematics, possessing interesting and significant problems of its own, and having connections with other parts of mathematics and with related fields such as theoretical computer science and mathematical economics.

Our interest in sets, however, stems from a different source. Over the course of the twentieth century, set theory has come to provide the basic language of mathematics. For example, many mathematicians maintain that any mathematical entity should be describable as a set. As a result, anyone wishing to study modern mathematics must become conversant with the vocabulary and concepts of set theory. Thus, we study set theory not as an end in itself, but as a language in which mathematical communication occurs.

A *set* is a collection of objects. Examples of sets are abundant: a set (or flock) of birds, a set (or pride) of lions, a set (or herd) of cattle. In everyday life, we deal with all sorts of sets of all sorts of things. In their daily work, mathematicians work with sets of numbers (e.g., real or rational), sets of functions (e.g., polynomial or trigonometric), sets of geometric objects (e.g., triangles, curves, or surfaces), or other sets of mathematical things.

If A is a set and x is one of the objects in the set A, then we write $x \in A$ and say that x is *an element of A* or x *is a member of A*. Thus, corn is an element of the set of vegetable crops grown in the United States. If x is not in A, then we write $x \notin A$.

How is a set specified? One method is simply to list the elements of the set. For instance, the numbers 2, 3, 4, and 5 constitute the set of integers between two and five inclusive. In the usual notation reserved for sets, this set is written $\{2, 3, 4, 5\}$. The braces, $\{\ \}$, are used to delineate the set.

It stands to reason that one can list all the elements of a set only when the set is rather small. In general, to describe a set, one states a property or condition that determines the elements that belong to the set.

Let P be a well-defined property of objects, mathematical or otherwise. For instance, P may be the property that a real number is positive or that a person is an American citizen. Then

$$\{x \mid P(x)\}$$

13

denotes the set of all elements having property P.

Example 1   The following sets of numbers are defined by properties.
(i) $A = \{x \mid x$ is a positive real number$\}$
(ii) $B = \{x \mid x$ is a positive rational number$\}$
(iii) $C = \{x \mid x$ is an integer$\}$

Just for emphasis note that in (i), $P(x)$ is the statement "x is a positive real number."

One of the most important sets in all of mathematics is the *empty set*, customarily denoted by the symbol $\emptyset$. By $\emptyset$ is meant the set having no elements. $\emptyset$ is also referred to as the *null set*.

Very often, we consider sets with just one element. Some examples are $\{1\}$, $\{$Don$\}$, $\{\emptyset\}$. These sets are referred to respectively as "singleton 1," "singleton Don," and "singleton $\emptyset$."

We summarize the ideas presented thus far and introduce the notion of a finite set in the following definition.

Definition 1   *Sets*

(i) A *set* is a collection of objects.
(ii) If x is in a set A, then we say that x *is an element of A* and write $x \in A$.
(iii) The *empty set*, $\emptyset$, is the set with no elements.
(iv) *Singleton x* denotes the set $\{x\}$ whose only member is x.
(v) A set is called *finite* if it is empty or its elements can be matched up precisely with elements of the set $\{1, 2, \ldots, n\}$, where n is some positive integer. A set that is not finite is called *infinite*.

For a finite set X, the integer n appearing in the definition is called the number of elements of X. In this case we write $X = \{x_1, x_2, \ldots, x_n\}$. Thus, the set of lower case letters of the English alphabet is a finite set with 26 elements. The set of positive integers is an infinite set.

Now we consider relationships that can exist between sets. The first concerns equality of sets, the second is the notion of containment.

Definition 2   *Set Equality*

Two sets A and B are *equal*, written $A = B$, when they have precisely the same elements. We write $A \neq B$ if A and B are not equal.

In other words, two sets A and B are equal when each element of A is an element of B and each element of B is an element of A. Symbolically, $A = B$ if and only if for each x, $x \in A$ if and only if $x \in B$ or $A = B \Leftrightarrow$ for all $x(x \in A \Leftrightarrow x \in B)$.

Definition 3   *Set Containment*

A set A is *contained in* a set B, written $A \subseteq B$, if each element of A is an element of B. If $A \subseteq B$, then we also say that A *is a subset of* B. We call A a proper subset of B if $A \subseteq B$ and $A \neq B$; in this case we write $A \subset B$. Finally we write $A \not\subseteq B$ if A is not a subset of B.

Example 2 $\{1, 2\} = \{2, 1\}$.

Example 3   For A, B, and C as defined in Example 1, $C \not\subseteq B$ and $B \subset A$.

Example 4 $\{x \mid x$ is a person living in the continental United States$\}$ $\subset \{x \mid x$ is a person living in the Western Hemisphere$\}$.

Let A and B denote sets. How does one show that $A \subseteq B$? According to Definition 3, one must demonstrate that each element of A is in B: For each x, if $x \in A$, then $x \in B$. In terms of defining properties, if $A = \{x \mid P(x)\}$ and $B = \{x \mid Q(x)\}$, then to show that $A \subseteq B$, we must argue that for each x the condition $P(x)$ implies the condition $Q(x)$. This type of proof, where we take an arbitrary $x \in A$, and show that $x \in B$, is called an *element-chasing proof*. Several examples of element-chasing proofs occur in this section. On the other hand, to show that $A \not\subseteq B$, we must show that an element exists that is in A and is not in B; this requires us to find or exhibit an element x such that $x \in A$ and $x \notin B$.

Example 5   Let $A = \{1\}$. The only subsets of A are $\emptyset$ and A (see Theorem 1, which follows).

Example 6   Let $A = \{\emptyset\}$ and $B = \{\emptyset, \{\emptyset\}\}$. We claim $A \subseteq B$, for if $x \in A$, then $x = \emptyset$ (note that A is not empty; it has one element and that element is the empty set). Then $\emptyset = x \in B = \{\emptyset, \{\emptyset\}\}$. Therefore, $A \subseteq B$. Notice that the elements of the sets A and B are themselves sets. Also observe that $A = \{\emptyset\} \in B$. Thus A is both a subset of B and an element of B. As it happens, many interesting sets in mathematics have other sets as their elements.

Our first theorem captures some basic facts about containment. Its proof provides our first example of an element-chasing argument.

Theorem 1   (i)   *For any set A, $\emptyset \subseteq A$.*
   (ii)   *For any set A, $A \subseteq A$.*
   (iii)   *If A, B, and C are sets with $A \subseteq B$ and $B \subseteq C$, then $A \subseteq C$.*

Proof   (i) We must demonstrate the truth of the statement: For each x, if $x \in \emptyset$, then $x \in A$. This conditional statement has a false antecedent, and hence is true. Therefore, $\emptyset \subseteq A$.

(ii) We must show: For each x if $x \in A$, then $x \in A$. But this conditional is a tautology (see Section 1, Exercise 8). Thus, $A \subseteq A$.

(iii) We must check the following statement: For each x if $x \in A$, then $x \in C$. By assumption, $A \subseteq B$; thus for each x if $x \in A$, then $x \in B$. However, since $B \subseteq C$ (by another assumption) and since $x \in B$, it follows that $x \in C$. Therefore, for each x if $x \in A$, then $x \in C$. Hence, $A \subseteq C$ whenever $A \subseteq B$ and $B \subseteq C$. ∎

According to Definition 2, two sets are equal when they have precisely the same elements. From this definition, the next result follows. The proof is left as an exercise.

Theorem 2   *Let A and B be sets. Then $A = B$ if and only if $A \subseteq B$ and $B \subseteq A$.*

To illustrate this theorem, we revive an exercise from high school algebra.

Example 7   Let us consider the solution set of the equality $x^2 - 2x - 4 = x - 6$. Let A denote this set:

$$A = \{x \mid x^2 - 2x - 4 = x - 6\}$$

What is A?

Let us suppose $x \in A$. What can we say about x? Now if $x \in A$, then
$$x^2 - 2x - 4 = x - 6,$$
or
$$x^2 - 3x + 2 = 0,$$
or
$$(x - 1)(x - 2) = 0,$$
which means that $x - 1 = 0$ or $x - 2 = 0$, which in turn implies that $x = 1$ or $x = 2$. Therefore, if $x \in A$, then $x = 1$ or $x = 2$, or, equivalently, $A \subseteq \{1, 2\}$.

It is now a simple matter to check that if $x \in \{1, 2\}$, then $x \in A$, i.e., that 1 and 2 are solutions of the equality $x^2 - 2x - 4 = x - 6$. Thus, $\{1, 2\} \subseteq A$.

Together, these remarks show that $A = \{1, 2\}$.

Representations of Sets

There are two standard methods of representing subsets of a set. These representations are helpful in investigating and understanding properties of subsets of a set. The first, pictorial in nature, is applicable to an arbitrary situation. The second, being digital, is primarily used to represent subsets of a finite set.

Venn Diagrams

Suppose A is a subset of a set X. Then A and X are represented as geometric figures in the plane as follows:

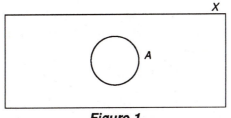

Figure 1

The exact shapes of A and X are irrelevant; what we have is visual representation of the general phenomenon of a subset of a set.

If B is another subset of X, then one might draw the following picture, at least under the assumption that A and B have elements in common:

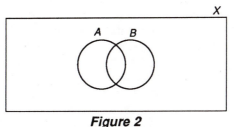

Figure 2

This geometric representation of subsets of a set provides a valuable aid to our intuition, assisting us in our investigation of properties of subsets and operations on subsets. These visual representations of subsets are called *Venn diagrams*.

Binary Sequences

Now to the digital representation of subsets of a finite set. Suppose n is a positive integer and let $X = \{x_1, \ldots, x_n\}$ be a finite set with n elements. Let A be a subset of x. The idea is to use a sequence of zeros and ones to keep track of the elements of A. Specifically, form a sequence of length n, each term of which is either 0 or 1.

Step 1. If $x_1 \in A$, then the first term of the sequence is 1; if $x_1 \notin A$, then the first term is 0.

Step 2. If $x_2 \in A$, then the second term of the sequence is 1; if $x_2 \notin A$, then the second term is 0.

$$\vdots$$

Continue in this fashion, until

$\vdots$

Step n. If $x_n \in A$, then the nth term of the sequence is 1; if $x_n \notin A$, then the nth term is 0.

Thus each subset $A \subseteq X$ corresponds to a sequence $a_1, a_2, \ldots, a_n$ where each entry a_i of the sequence is either 0 or 1. It is customary and causes no ambiguity to suppress the commas and to write the terms consecutively without spaces between them; hence, A corresponds to $a_1 a_2 \cdots a_n$. An example will illustrate.

Example 8   Let $X = \{1, 2, 3, 5\}$. If $A = \{1, 5\}$, then the sequence corresponding to A is 1001. The sequence corresponding to the empty set, $\emptyset$, is 0000; the sequence determined by X itself is 1111. Question: What subset is assigned to the sequence 1101? Note that all sequences considered in this example have four digits since the set X has four elements.

In general, a sequence of length n, $a_1 \cdots a_n$, where each a_i is either 0 or 1 is called a *binary string of length n*. Therefore, each subset of a set with n elements can be represented as a binary string of length n. (In general, a *string* is a sequence of symbols chosen from a fixed finite set called the *alphabet*; the word "binary" refers to the fact that in this case the alphabet has only two elements, 0 and 1.)

The binary representation of subsets of a finite set is extremely useful. For example, it facilitates the manipulation of sets on a digital computer. Also, many set operations can be analyzed from this viewpoint. Finally, as the exercises at the end of this unit will show, many counting problems having to do with finite sets can be approached successfully using the binary representation.

Set Operations

Ever since grade school we have worked with the arithmetic operations of addition, subtraction, and multiplication. These operations possess a common feature. Addition of real numbers provides a way of associating to each pair of real numbers, a, b, a real number $a + b$. Similarly, subtraction and multiplication are procedures that assign a real number to each pair of real numbers. Subtraction assigns $a - b$ to the pair a, b, while multiplication associates to a and b the number ab. In general, each of these operations assigns to every pair of real numbers another real number. We now discuss operations on sets—union, intersection, and set difference—that are analogous to these operations on real numbers. Each of these operations works by assigning to each pair of sets another set.

Definition 4 *Union*

   Let A and B be sets. The *union of A and B*, written $A \cup B$, is the set

$$A \cup B = \{x \mid x \in A \text{ or } x \in B\}.$$

In words, the union of A and B is the set of elements that are in at least one of the two sets A and B.

Definition 5 *Intersection*

   Let A and B be sets. The *intersection of A and B*, written $A \cap B$, is the set

$$A \cap B = \{x \mid x \in A \text{ and } x \in B\}.$$

Thus, $A \cap B$ is the set of elements common to both A and B.

Example 9   Let $A = \{1,2,3\}$, $B = \{3,4,5\}$, and $C = \{5,6,7\}$. Then $A \cup B = \{1,2,3,4,5\}$, $A \cap B = \{3\}$, $A \cup C = \{1,2,3,5,6,7\}$, and $A \cap C = \emptyset$.

The union of two sets is accomplished by combining or putting together the elements of the individual sets into a new set. Roughly speaking, then, taking the union of two sets is analogous to adding two real numbers. We ask: Does subtraction of real numbers have a set theoretic analog? Thinking of subtraction as the taking away or removal of numbers or objects, we are led to the next definition.

Definition 6 *Set Difference*

   Let A and B be sets. The *set difference of* or *difference of A and B*, is defined to be

$$A - B = \{x \mid x \in A \text{ and } x \notin B\}.$$

(Some authors use the symbol $A \backslash B$ in place of $A - B$.)

Example 10   Let $\mathbf{Z}$ denote the set of integers: $\mathbf{Z} = \{0, \pm 1, \pm 2, \pm 3, \ldots\}$.
(i) Let $E = \{z \in \mathbf{Z} \mid z \text{ is an even integer}\}$. Then $\mathbf{Z} - E = \{z \in \mathbf{Z} \mid z \text{ is an odd integer}\}$.
(ii) Let $A = \{1,2,3\}$ and $B = \{2,3,4\}$. Then $A - B = \{1\}$ and $B - A = \{4\}$.

Notice that in the definition of the set E, we restricted the domain of the variable z to the set of integers.

The following concept, derived from the operation of set difference, arises frequently.

Definition 7 *Set Complement*

Let A be a subset of a set U. The *complement of A in U* is the set $U - A$.

If $A \subseteq U$ and $A \subseteq V$, then the sets $U - A$ and $V - A$ may be unequal, in which case the complement of A in U will differ from the complement of A in V. However, if we fix the set U throughout the discussion, then we call $U - A$ the *complement of A* and denote it by A^c.

Example 11 Let U be the set $\mathbf{Z}$ of integers. Let $\mathbf{N}$ denote the set of nonnegative integers: $\mathbf{N} = \{0, 1, 2, 3, \dots \}$. Then $\mathbf{N} \subseteq \mathbf{Z}$ and $\mathbf{N}^c = \{x \in \mathbf{Z} \mid x < 0\}$. If $E = \{x \in \mathbf{Z} \mid x$ is even$\}$, then $E^c = \{x \in \mathbf{Z} \mid x$ is odd$\}$.

Representations of Operations

We now consider the interpretations of the set operations of $\cup$, $\cap$, $-$, and ^{c} in terms of our methods of representing sets—Venn diagrams and binary sequences.

Suppose A and B are subsets of a set U. Then Venn diagrams for $A \cup B$ and $A \cap B$ are presented as shown in Figure 3.

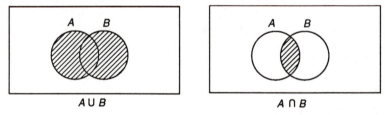

Figure 3

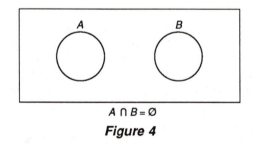

Figure 4

As is evident from either the definitions or the Venn diagram, $A \cap B$ is a subset of both A and B, which in turn are subsets of $A \cup B$. The proofs of these observations are left as exercises.

In drawing A and B as we have, we leave open the possibility that $A \cap B = \emptyset$. If, however, we wish to depict unambiguously the case that $A \cap B = \emptyset$, then we represent A and B as shown in Figure 4.

The Venn diagrams for the set difference, $A - B$, and the complement, A^c, are depicted in Figure 5.

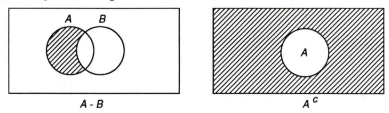

Figure 5

We next interpret the set operations of intersection, union, and complement in terms of binary sequences. Let $U = \{x_1, \ldots, x_n\}$ be a finite set with n elements. Let A and B be subsets of U and let $a_1 \cdots a_n$ and $b_1 \cdots b_n$ be the binary sequences corresponding to A and B, respectively. How can the binary sequences of $A \cap B$, $A \cup B$, and A^c be determined from the binary sequences of A and B?

Let us consider an example. Let $U = \{1, 2, 3, 4\}$, $A = \{1, 2\}$, and $B = \{2, 3\}$. Then 1100 and 0110 are the binary sequences determined by A and B, respectively. Now $A \cap B = \{2\}$, $A \cup B = \{1, 2, 3\}$, and $A^c = \{3, 4\}$, and the binary sequences corresponding to these sets are 0100, 1110, and 0011, respectively. How are each of these sequences related to 1100 and 0110?

Perhaps the sequence belonging to A^c is the simplest to describe. The sequence 0011 (which is determined by A^c) is obtained from 1100 (which corresponds to A) by changing each 0 in 1100 to 1 and each 1 in 1100 to 0. This observation generalizes to an arbitrary subset A of an arbitrary finite set U. Let $a'_1, \ldots, a'_n$ be the binary sequence of A^c. (Recall that the binary sequence of A is $a_1 \cdots a_n$). Then

$$a'_i = \begin{cases} 1 & \text{if } a_i = 0 \\ 0 & \text{if } a_i = 1. \end{cases}$$

Quick proof: If $a_i = 0$, then $x_i \notin A$; thus $x_i \in A^c$ and $a'_i = 1$. If $a_i = 1$, then $x_i \in A$; hence $x_i \notin A^c$ and $a'_i = 0$.

What about the binary sequence for $A \cap B$? Let $c_1 \cdots c_n$ be the binary sequence determined by $A \cap B$. Then for $1 \le i \le n$, c_i can be expressed in terms of a_i and b_i as follows:

$$c_i = \begin{cases} 1 & \text{if } a_i = 1 = b_i \\ 0 & \text{if } a_i = 0 \text{ or } b_i = 0 \end{cases}$$

for, if $a_i = b_i = 1$, then $x_i \in A \cap B$; hence $c_i = 1$. If $a_i = 0$ or $b_i = 0$, then either $x_i \notin A$ or $x_i \notin B$; therefore $x_i \notin A \cap B$ and $c_i = 0$. Notice that the relationship among a_i, b_i, and c_i can be expressed more concisely as $c_i = a_i \cdot b_i$. In other words, the ith entry of $c_1 \cdots c_n$ is the product of the ith entries of $a_1 \cdots a_n$ and $b_1 \cdots b_n$.

Let $d_1 \cdots d_n$ denote the binary sequence of $A \cup B$. We leave as an exercise the proof that

$$d_i = \begin{cases} 1 & \text{if } a_i = 1 \text{ or } b_i = 1 \\ 0 & \text{if } a_i = 0 = b_i. \end{cases}$$

The Set of Real Numbers

The most important set in all of mathematics is (arguably) the set of real numbers. The set of real numbers and its subsets are the starting points for several areas of study within mathematics including analysis (calculus and its generalizations), algebra, geometry, and topology. In the school mathematics curriculum, the real numbers are featured in courses such as algebra, trigonometry, and calculus. Because the real number system is a cornerstone of mathematics, a working knowledge of the real numbers is necessary for anyone wishing to do or to use mathematics.

At this point, our discussion of the real numbers will be informal. We present a visual representation of the real numbers as points on a line. In Unit II we discuss an axiomatic treatment of the real numbers.

Perhaps the simplest way of envisioning the real number system is geometrically. Imagine a straight line stretching without bound in both directions. The idea is to match a real number to each point on this line. Henceforth this line is referred to as a *number line*.

We begin by marking a point on the line and labeling it 0. Next choose a unit distance, mark off one unit to the right of 0, and label this point 1 (Figure 6).

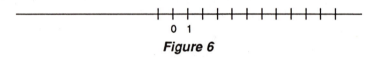

Figure 6

The point one unit to the right of 1 is labeled 2. The set of numbers that label the points obtained by starting with 0 and repeatedly taking points one unit to the right of the previously labeled point as in Figure 7 is called the set of *natural numbers* and is denoted by **N**. The members of the set **N** are the numbers $0, 1, 2, 3, \ldots$: $\mathbf{N} = \{0, 1, 2, 3, \ldots\}$.

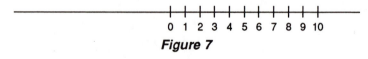

Figure 7

If we mimic this geometric construction of the left side of 0, then we obtain numbers that are labeled -1, -2, -3, ..., as shown in Figure 8.

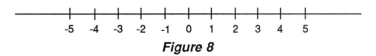

Figure 8

These numbers taken together with the set of natural numbers form a new collection, called the set of *integers* and denoted by **Z** (**Z** is an abbreviation for the German word for numbers, *Zahlen*). Thus **Z** is set $\{0, \pm 1, \pm 2, \pm 3, \ldots\}$. The integers $1, 2, 3, \ldots$ are called the *positive integers* and the numbers $-1, -2, -3, \ldots$ are called the *negative integers*. From the definition of **Z**, it follows that any natural number is an integer, i.e., $\mathbf{N} \subset \mathbf{Z}$.

Next we define the set of *rational numbers*, denoted by **Q**. A rational number can be represented as a ratio or quotient of integers, m/n, where $n \neq 0$. In set notation, $\mathbf{Q} = \{m/n \mid m, n \in \mathbf{Z}, n \neq 0\}$. Recall that an element of **Q** can be represented as a ratio of integers, i.e., as a *fraction*, in more than one way. For example, $1/2 = 2/4 = 3/6$. In general the fractions m/n and m'/n' are *equivalent*, written $m/n = m'/n'$ if $mn' = m'n$.

Geometrically the numbers m/n are obtained from the following process: We consider two cases, the first being when n is a positive integer. Divide the segment from 0 to 1 into n equal pieces. Take the first of these segments, the one whose left end is the point 0. If m is positive, then reproduce this segment m times to the right of 0. The right endpoint of the segment so obtained corresponds to m/n. If m is negative, then reproduce the segment $-m$ times to the left of 0 to obtain m/n. (See Figure 9.) The second case occurs when n is negative. In this case we correspond m/n to the point $(-m/(-n))$ as constructed in the previous case. Note that any integer is also a rational number. It is clear geometrically that any integer n corresponds to the fraction $n/1$. Thus $\mathbf{Z} \subset \mathbf{Q}$.

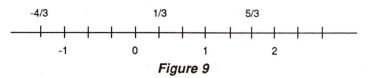

Figure 9

Clearly this process allows us to attach numerical labels to many points on the line. In fact, if a and b are points with rational labels, then between a and b there lies a point corresponding to a rational number. Thus the rational numbers are not separated geometrically in the way the integers are; the rationals are distributed quite densely along the number line. One might even ask: Does every point on the line correspond to a rational number? Is every point on the line obtained by the geometric construction described in the previous paragraph? Many early Greek mathematicians, especially the followers of the mathematician Pythagoras, believed, primarily on philosophical and religious grounds, that indeed every point on the line is matched to a rational number. The idea that all distances are rational (*commensurable* was the word the Greeks used) was central to the

Pythagorean conception of the order of the universe. Thus, they were dismayed by the discovery of Hippasus, himself a Pythagorean, who lived in the 5th century B.C., that there are distances, i.e., points on the line, that are not rational. Specifically, Hippasus showed that $\sqrt{2}$ is irrational, i.e., not rational. (We give Hippasus's proof that $\sqrt{2}$ is irrational in Section 5). Legend has it that Hippasus announced that $\sqrt{2}$ is not rational while on a ship with other Pythagoreans and was rewarded for his blasphemous discovery by being tossed overboard.

The fact that not all points on the line correspond to rational numbers opens the way for the introduction of even more numbers. We show how to represent each point on the line, rational or irrational, by the so-called decimal expansion. Given a point on the line, first find the integer point immediately to its left; for x as in Figure 10, this

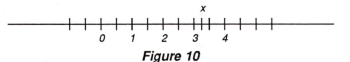

Figure 10

integer is 3. Next on the segment from 3 to 4, find the one-tenth mark immediately to the left of x. In Figure 11, this is the number

Figure 11

$31/10 = 31/10$, which we write in decimal representation as 3.1. Repeat this procedure for the segment between $31/10$ and $32/10$. In Figure 12, x lies between the points $314/100$ and $315/100$, which are

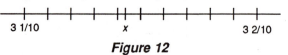

Figure 12

written in decimal form as 3.14 and 3.15, respectively. Continuing this process, we obtain the decimal representation of x:

$$x = 3.1415926535\ldots.$$

Each of the rational numbers 3, 3.1, 3.14, 3.141, 3.1415, ... lies to the left of x on the number line. This procedure can be repeated for any arbitrary point x on the number line. Thus, each point on the line is assigned a decimal representation. Most numbers have unique decimal expansions: If $x = .a_1a_2a_3\ldots = .b_1b_2b_3\ldots$ are two decimal expansions of x, then $a_1 = b_1$, $a_2 = b_2$, $a_3 = b_3,\ldots$ with the exception of numbers terminating in 9s or 0s. For instance, $.12999\ldots = .13000\ldots = .13$.

The set of such decimal expansions is called the set of real numbers and is denoted by **R**. Observe that any rational number is a real number, i.e., $\mathbf{Q} \subset \mathbf{R}$. In fact one can show that a real number x is rational if and only if the decimal expansion of x terminates (such as $1/8 = .125000\ldots$) or repeats (such as $12/99 = .121212\ldots$).

We have then two informal perspectives from which we can regard the real numbers. First, from the geometric perspective we view the set of real numbers as an indefinitely extended line on which each point corresponds to a unique real number. This geometric viewpoint has several benefits. For example, it provides us with a way of grasping the real numbers as a whole. Often certain properties of the real numbers (for instance, the transitivity property of $<$ proved in Section 6) are obvious when considered geometrically. Second, we have the decimal representation of a real number. Each real number has a decimal expansion. This expansion is unique except for those real numbers whose decimal expansion can be represented from some point on using only the digit 9. The decimal representation is useful whenever precise calculations with real numbers are necessary.

EXERCISES §2

1. List the elements in each of the following sets:
   (a) $\{x \in \mathbf{R} \mid x^2 - 10x + 7 = 0\}$.
   (b) $\{x \in \mathbf{R} \mid x^2 - 8x + 5 = 0\}$.
   (c) $\{x \in \mathbf{R} \mid 1/x + 1/(x+1) = 2\}$.
   (d) $\{x \in \mathbf{R} \mid x^3 - 3x = 0\}$.
   (e) $\{x \in \mathbf{R} \mid x^2 + 4x + 5 = 0\}$.
2. (a) Show: $\{1, 2, 5\} = \{5, 2, 1\}$
   (b) Show: $\{\{0\}, \{0, 1\}\} \neq \{\{0\}, \{1\}\}$
   (c) Show: $\{a, a\} = \{a\}$
   (d) Show: $\{\{a\}, \{\{a\}\}\} \neq \{\{a\}\}$.
   (e) Show that $\{0\} \in \{\{0\}, 1\}$ but $\{0\} \not\subset \{\{0\}, 1\}$.
3. (a) Find real numbers a and b such that $\{x \in \mathbf{R} \mid x^2 - x - 2 < 0\} = \{x \in \mathbf{R} \mid a < x < b\}$.
   (b) Find real numbers a and b such that $\{x \in \mathbf{R} \mid |x + 1| < 5\} = \{x \in \mathbf{R} \mid a < x < b\}$.
   (c) Show that $\{x \in \mathbf{R} \mid (x - 1/2)(x - 1/3) < 0\} \subseteq \{x \in \mathbf{R} \mid 0 < x < 1\}$
4. Prove Theorem 2.
5. (a) Suppose $B \subseteq A$. Draw an appropriate Venn diagram for this situation.
   (b) Suppose A and B have no elements in common. Draw an appropriate Venn diagram.
6. Let $A = \{1, 2, 3, 4, 5\}$. List the binary sequences corresponding to each of the following subsets of A.
   (a) $\emptyset$ (b) A (c) $\{1, 2, 3\}$ (d) $\{1, 3, 5\}$ (e) $\{2, 4, 3\}$.
7. List all subsets of (a) $\{1, 2\}$,   (b) $\{1, 2, 3\}$, and   (c) $\{1, 2, 3, 4\}$. Next to each subset, give its corresponding binary sequence. Note: In solving (b) (and, respectively, (c)), try to use your list from (a) (respectively, (b)) in a systematic way.

8. A set A is called *full* if any element of A is also a subset of A. In other words, A is full if $x \in A$ implies $x \subseteq A$.
   (a) Show that $\{\emptyset\}$ is full.
   (b) Find a full set having exactly two elements.
   (c) Find a full set having exactly three elements.
   (d) Generalize your findings from (b) and (c).
9. The aim of this exercise is to arrive at a conjecture for the number of subsets of a set with n elements. Form a table as follows:

n	1	2	3	4
B_n				
S_n				

Here n denotes the number of elements of a finite set X, S_n the number of subsets of X, and B_n the number of binary strings of length n. In parts (c) and (d) you are asked to pinpoint the general pattern that is suggested in the table.
   (a) How many binary strings are there of length 1? length 2? length 3? length 4?
   (b) How many subsets are there in a set with one element? two elements? three elements? four elements? (See Exercise 7.)
   (c) For any arbitrary positive integer n, how many binary strings of length n are there?
   (d) Let n be a positive integer and let X be a set with n elements. How many subsets does X contain?
10. Let $X = \{x_1, \ldots, x_n\}$. Suppose that $A \subseteq X$ and that A is represented by the binary sequence $a_1 a_2 \ldots a_n$.
   (a) How can the number of elements of A be determined from $a_1 a_2 \ldots a_n$?
   (b) Which sequence corresponds to $\emptyset$?
   (c) Which sequence corresponds to X?

Section 3
PREDICATES AND QUANTIFIERS

While the propositional calculus is an essential tool in mathematical work, not all mathematical statements are propositions. For example, the sentence "$x^2 + 2x - 3 = 0$" is neither true nor false as it stands. If $x = 1$, then the sentence becomes a true proposition; if $x = 2$, then the resulting proposition is false. In the sentence "$x^2 + 2x - 3 = 0$," the symbol "x" represents a number, perhaps an arbitrary real number or perhaps a number chosen from a restricted collection of real numbers. In order to deal with sentences involving a symbol or symbols that represent an object or objects chosen from a set of objects, we need a new language. This language is provided by the predicate calculus.

The predicate calculus is concerned with arguments and inferences involving sentences. In addition to the logical connectives used in propositional calculus, the predicate calculus involves *variables*, *predicates*, and *quantifiers*. We now define and illustrate each of these concepts.

Definition 1 *Variable*
A *variable* is a symbol representing an unspecified object that can be chosen from some set of objects.

Consider the statements:
1. If x is a nonzero real number, then $x^2 > 0$.
2. If T is a triangle in the plane, then the sum of the interior angles of T is 180°.

In the first statement, x represents a number chosen from the set of all nonzero real numbers; x can be *any* nonzero real number. In the second statement, the symbol T represents any triangle in the set of all triangles in the plane.

Definition 2 *Predicate*
A *predicate* is a sentence $P(x_1, \ldots, x_n)$ involving variables $x_1, \ldots, x_n$ with the property that when specific values from a given set are assigned to $x_1, \ldots, x_n$, the resulting statement is either true or false.

For example, the sentence "x is less than y" is a predicate that is true when $x = 2$ and $y = 3$ but is false when $x = 3$ and $y = 2$.

One property of a variable that must be specified in order to avoid ambiguity is its *scope*, i.e. the set of objects, a member of which the variable represents or symbolizes. For example, to emphasize the fact that the symbol x in statement 1 can be an arbitrary nonzero real, that statement can be rewritten as:

3. For every real number x, if $x \neq 0$, then $x^2 > 0$, or
4. For every x, if x is a nonzero real number, then $x^2 > 0$.

The phrase "for every" prescribes the scope of the variable x. Because of its frequent appearance in logic and mathematics, this phrase has a name.

Definition 3   *Quantifiers*

The phrase "for every" is called a *universal quantifier*. The phrase "there exists" is called an *existential quantifier*.

Statements 1, 3, and 4 are interpreted to mean that for each choice of a nonzero real number x, it is the case that $x^2 > 0$. Statement 2 can be written with a universal quantifier as:

5. For every T, if T is a triangle in the plane, then the sum of the interior angles of T is 180°.

In general, a statement involving a universal quantifier has the form "For every x, $P(x)$" where $P(x)$ is a sentence asserting that x possesses some property. In logic texts (and in some mathematics books) the symbol $\forall$ is used to represent the phrase "for every." With this notation the sentence "For every x, $P(x)$" becomes "$\forall x$, $P(x)$." We will rarely use the symbol "$\forall$" after this section. Finally, the phrase "for every" is often written as "for each" and "for all." We will use these three terms interchangeably.

The basic form of a statement involving an existential quantifier (with a single variable) is "There exists x such that $P(x)$." The symbol "$\exists$" is often used as an abbreviation for "There exists." The phrases "some x," "for some x," and "for at least one x" also denote existential quantifications. For example, the statement

6. There exists x such that x is a nonzero real number, and $x^2 > 0$.

means that for *at least one* choice of a nonzero real number x, it is the case that $x^2 > 0$. Assertion 6 can also be phrased as:

7. $x^2 > 0$ for some nonzero real number x.

It is almost always the case that a clearer understanding of a statement can be obtained by rewriting it using variables and quantifiers. Here are some examples:

8. Every real number is either positive, negative, or zero.
8'. For every x, if x is a real number, then either x is positive, x is negative, or x is zero.
9. Some triangles are isosceles.

9'. There exists T such that T is a triangle and T is isosceles.

10. The sum of two rational numbers is a rational number.

10'. For every x and every y, if x and y are rational, then $x + y$ is rational.

More examples are provided in the exercises. In many cases, ambiguity arises when proper quantification is omitted from a statement: Consider the equation $x + 1 = 0$. The quantification on x is missing. By supplying quantification we obtain either

        (i)   For every x, $x + 1 = 0$.

or   (ii) There exists x such that $x + 1 = 0$.

But even these forms are not complete, since we have not specified the set from which x is to be chosen. For example, if x is allowed to be an arbitrary real number, then (i) is false while (ii) is true. But if x is taken only from the set of positive numbers, then both (i) and (ii) are false. On the other hand, if x is taken from the set of numbers whose only member is the number -1, then both (i) and (ii) are true. Thus, the moral is: In any statement involving variables, be sure that each variable is properly quantified and remember the scope of each variable.

Frequently, it is necessary to form the negation of a statement involving variables and quantifiers. Such negations can be tricky. Let us look at a few examples.

Example 1   Form the negation of the statement: "if x is a nonzero real number, then $x^2 > 0$."

The safest way to begin is to add the appropriate quantification: "for every x, if x is a nonzero real number, then $x^2 > 0$." Perhaps the most prudent next step is to state the negation by adding the phrase "it is not the case that" at the beginning of the given statement:

It is not the case that for every x, if x is a nonzero real number, then $x^2 > 0$.

Now what does it mean to say that "it is not the case that for every x, $P(x)$?" Evidently, this means that it is the case that for some x, the negation of $P(x)$ holds: There exists an x such that $P(x)$ does not hold. Thus, we must negate the statement "if x is a nonzero real number, then $x^2 > 0$." But the implication $P \Rightarrow Q$ is logically equivalent to "(not-P) or Q," and hence "not-$(P \Rightarrow Q)$" is logically equivalent to "not-[(not-P) or Q]," which is logically equivalent to "P and not-Q." Therefore, we can state the negation of the original statement as:

There exists x such that x is a nonzero real number and $x^2 \not> 0$.

We can generalize this example as follows: The negation of a statement of the form "for every x, $P(x)$" is the statement "there exists x such that not-$P(x)$". On the other hand, the negation of "there exists x such that

$Q(x)$" is the statement "for each x, not-$Q(x)$" (i.e., for each x, it is not the case that $Q(x)$ holds). We summarize these remarks:

Statement	Negation
For every x, $P(x)$.	There exist x such that not-$P(x)$.
$(\forall x \; P(x))$	$(\exists x(\text{not-}P(x)))$
There exists x such that $Q(x)$.	For every x, not-$Q(x)$.
$(\exists x \, Q(x))$	$(\forall x(\text{not-}Q(x)))$

The statements $P(x)$ and $Q(x)$ might themselves be quite involved and difficult to negate. The best way to proceed is to write $P(x)$ and $Q(x)$ in terms of logical connectives and use the rules for negation discussed earlier in this section. This procedure was followed in Example 5. Let us illustrate by negating the quantified statement: $(\forall x)(P(x) \wedge Q(x))$. If this statement does not hold, then there exists x such that it is not the case that $P(x) \wedge Q(x)$ holds. Therefore, the negation of the given statement is $(\exists x)(\text{not-}(P(x) \wedge Q(x)))$, which becomes $(\exists x)((\text{not-}P(x)) \vee (\text{not-}Q(x)))$.

Example 2   Negate the statement: There exists a real number whose square is -1.

First, let's write this statement using a variable. Here is a reasonable rendition:

$$\exists x \text{ such that } x \in \mathbf{R} \text{ and } x^2 = -1.$$

Thus the negation of this statement is

$$\forall x \text{ it is not the case that } x \in \mathbf{R} \text{ and } x^2 = -1.$$

We can modify this statement to obtain

$$\forall x \text{ either } x \notin \mathbf{R} \text{ or } x^2 \neq -1.$$

When making this statement, we probably are considering x to be a real number. Thus we can restate the last assertion as:

$$\forall x \in \mathbf{R}, \quad x^2 \neq -1.$$

A final word about quantification. One often encounters two or more quantifiers in a given statement. Here are some examples (x and y are real numbers).

For all x and for all y, $x + y = y + x$;

There exists (a number) 0 such that for all y, $0 + y = y$; and

For all x there exists y such that $x + y = 0$.

It is important to observe that the order of the quantifiers is crucial. For instance, the statements "for all x there exists y such that $x + y = 0$," and "there exists y such that for all x, $x + y = 0$" have entirely different meanings. The first asserts that for any given real number x, a real number y can be found for which the equation $x + y = 0$ holds. The number y depends on the given number x. The second statement claims that there exists a single number y for which the equation $x + y = 0$ holds for all numbers x. Thus care must be taken when both an existential and a universal quantifier appear in the same sentence.

EXERCISES §3

In exercises 1–11 each of the statements is a theorem of calculus or precalculus. In every case (a) write the contrapositive of the statement, and (b) write the converse and state if the converse is true.

1. For all real numbers a and b, if $a > 0$ and $b > 0$, then $a \cdot b > 0$.
2. If $\sum a_n$ is a convergent infinite series, then $\lim\limits_{n \to \infty} a_n = 0$.
3. If f is continuous on $[a, b]$, then $\displaystyle\int_a^b f(x)\, dx$ exists.
4. If f is constant on $[a, b]$, then $f'(x) = 0$ for all x in (a, b).
5. Let T be a triangle having sides a, b, c with c being the longest side. If T is a right triangle, then $c^2 = a^2 + b^2$.
6. If $\sum |a_n|$ converges, then $\sum a_n$ converges.
7. If f is continuous on $[a, b]$, then f attains a maximum value on $[a, b]$.
8. If f has a local (relative) maximum at the real number a or a local (relative) minimum at a, then $f'(a) = 0$ or $f'(a)$ does not exist.
9. Suppose for every positive integer n, a_n and b_n are real numbers such that $0 \le a_n \le b_n$. If $\sum a_n$ diverges, then $\sum b_n$ diverges.
10. With the same hypotheses on a_n and b_n as in 9, if $\sum b_n$ converges, then $\sum a_n$ converges.
11. If T is an equilateral triangle, then T is an equiangular triangle.

In exercises 12–21 (a) write each of the statements using variables and quantifiers and (b) negate each statement.
12. Some integers are perfect squares.
13. Every rational number is a real number.
14. The product of two rational numbers is rational.
15. The cube root of any integer is irrational.
16. The square root of any irrational number is irrational.
17. No solution of $ax = b$ where a and b are arbitrary integers is irrational.
18. The derivative of a constant function is 0.
19. If x and y are irrational, then $x + y$ is irrational.
20. $|x + 7| = 3$ if and only if $x = -10$ or $x = -4$.

21. Form a negation of each of the following quantified statements:
    (a) $(\forall x)(P(x) \vee Q(x))$.
    (b) $(\forall x)(P(x) \Rightarrow Q(x))$.
    (c) $(\forall x)(P(x) \Leftrightarrow Q(x))$.
    (d) $(\exists x)(P(x) \wedge Q(x))$.
    (e) $(\exists x)(P(x) \vee Q(x))$.
    (f) $(\exists x)(P(x) \Leftrightarrow Q(x))$.

UNIT 2
PROVING
THEOREMS

If we look upon the decisive moment in the development of mathematics, the moment when it took its first step and when the ground on which it is based came into being—I have in mind logical proof—...
            I. R. Shafarevich

The object of mathematical rigor is to sanction and legitimate the conquests of intuition ...
            Jacques Hadamard

A good proof is one which makes us wiser.
            Yu. Manin

The opinions presented by Shafarevich and Hadamard represent two extreme positions on the role of proof and rigor within mathematics and indeed on the nature of mathematics itself. Both Hadamard (1865–1963) and Shafarevich (1923–   ) have made many significant contributions to mathematics; and so, to gain a better understanding of the place that proof occupies within mathematics, we should look carefully at each of their comments.

According to Shafarevich, the existence of proofs of mathematical statements distinguishes mathematics among all intellectual endeavors. In other scientific fields, investigators can at best propose conjectures about the essence of things and attempt to confirm these conjectures with experimental data or observation of natural phenomena. In fact, most scientific theories "explain" the events of nature in terms of other scientific or mathematical theories. For example, geologists account for earthquakes via the theory of plate tectonics, and physicists describe the motion of planets in our solar system in terms of a system of differential equations. However, due to the complex and haphazard way of the world, the fallibility of human observational methods, or many other factors, the possibility always exists that a scientific theory will have to be abandoned or modified because of empirical evidence.

In mathematics the situation is different. Mathematicians working in a particular field

1. establish certain undefined terms,
2. agree on certain fundamental rules, i.e., the axioms or postulates, relating the undefined terms,
3. introduce some auxiliary concepts, called definitions,
4. deduce statements, called (depending on their importance) theorems, propositions, lemmas, or corollaries relating the items that are presented in the axioms or definitions.

A particular statement becomes a theorem (proposition, lemma) when it is justified or proved. The proof of a statement consists of a sequence of statements, beginning with an axiom or previously established statement, and proceeding through a sequence of statements where each is deducible using rules of logic from axioms, previous theorems, or earlier statements in the sequence. Without the concept of proof, mathematics would be another empirical science. But with this notion, mathematics becomes *the* deductive science, hence its uniqueness and its justification.

Hadamard, on the other hand, feels the essence of mathematics lies in its discovery of interesting and striking relationships among the objects of the mathematician's fancy—numbers, functions, curves, surfaces, groups, Boolean algebras, etc. Once the theorems are articulated, proofs must be given, but only to legalize the theorems. Their truth is visible to the experienced mathematician before the proofs are given. Some philosophers of mathematics extend this position to entire mathematical theories: They argue that in practice the axioms of a particular mathematical discipline are actually created last, as a postscript to the discovery of the theorems of the subject, even though in any formal exposition of the discipline (in a textbook or research paper), the axioms head the parade with the theorems and their proofs marching faithfully behind. From this viewpoint, mathematics is closely akin to the sciences. Results are discovered, then justified and organized into a comprehensible framework.

The debates concerning the nature of mathematics and the role of proof within mathematics rage to this day. Of course, issues of this sort may never be resolved to the satisfaction of all parties. However, discussion such as the one outlined in the previous paragraph help us understand the purpose, practice, and essence of mathematics. From them we obtain a more detailed portrait of mathematics and a clearer fix on its position relative to the other intellectual activities of humans.

On the basis of the preceding discussion, what can we conclude about the role of proof in mathematics?

On the one hand no one familiar with mathematics will deny the importance of rigor and proof. Experience has convinced every mathematician that any assertion must be supported by a rigorous argument. Countless novice, apprentice, and master mathematicians have attempted to prove an intuitively obvious "theorem" only to watch it disintegrate into a false statement. Thus, every mathematician must become proficient with proofs. He or she must learn to read mathematical proofs critically as they are presented in texts or research papers and must develop the ability to write valid proofs. Proofs are indeed one of the focal points of mathematical activity; hence they need to be mastered by any student of formal mathematics.

On the other hand, activities such as experimentation, discovery, and reasoning by analogy are important aspects of a mathematician's experience. Mathematicians investigate abstract phenomena in search of order and pattern. Employing a combination of informal reasoning and experimentation, they conjure up conjectures that describe the patterns that they observe. At this point either they

(i) prove their conjectures formally,
(ii) disprove them by constructing counterexamples, in which case the process of investigation and conjecture repeats itself, or
(iii) get stuck, in which case either they
    (a) gather numerical evidence,
    (b) modify the problem somehow in the hope of obtaining a solvable problem, or
    (c) curse in frustration.

From this point of view, mathematics encompasses a broad range of activities, requiring mathematicians to guess and to prove, to hypothesize and to criticize.

Perhaps now, we see that the line between formal and informal mathematics begins to blur. Proof itself often acts as an aid for discovery. Using methods of proof to be discussed in this section, a mathematician can investigate several possible avenues toward a solution and eliminate those outcomes that cannot actually arise. At times, a mathematician will set out to prove a particular statement and will choose what appears to be a natural method of proof only to find that the proof forces a modification of the statement being proved. What arises is a kind of dialogue between two faces of mathematics, one being that of rigor, logic, and criticism, the other that of intuition, creativity, and discovery. In this dialogue, proof serves as a tool for both investigation and verification.

Perhaps now the wisdom of Manin's words quoted at the start of this section becomes apparent. (Manin, like Shafarevich, a contemporary Russian mathematician, has conducted important research in areas of pure mathematics and mathematical physics.) A good proof makes us wiser in several ways: The proof convinces us of the validity of the statement that it demonstrates. Once we have secured a proof, we can be certain that the result is valid and can be used whenever appropriate. The proof helps us understand the statement by providing insight into its nature. For instance, sometimes the proof of a theorem will "require" results and techniques that are not suggested by the statement of the theorem. Finally, the proof often directs us toward the result itself. Thus, a good proof confirms, clarifies, and creates mathematical results.

In this unit, we discuss the structure and mechanics of mathematical proofs. We describe and illustrate several kinds of mathematical proofs. Our overall goal is to become familiar with various proof techniques and to acquire a feeling for the appropriateness of a given proof technique in a particular situation. Admittedly, the latter skill is not readily acquired. Most mathematicians probably agree that it develops from experience with reading and writing proofs. Nevertheless, the presentation of proof techniques and the providing of reasons for choosing a specific proof method can help you to choose a reasonable proof strategy and to develop this strategy into a complete proof.

In this unit, five general and commonly used proof techniques are discussed:
- (i) direct proof
- (ii) proof by contrapositive
- (iii) proof by contradiction
- (iv) mathematical induction
- (v) case analysis.

Our goal is to describe the essence of each method and to illustrate each one in action. In addition, we indicate situations in which a given method might be especially appropriate. Before diving into a discussion of proof techniques, however, it is prudent to ask a simple question: What is the structure of the assertions that one encounters in mathematics?

Most statements of mathematical theorems are of the "If-then" form:
1. If a function f is differentiable at a, then f is continuous at a.
2. If a, b are real numbers and $ab = 0$, then either $a = 0$ or $b = 0$.
3. If G is a finite simple group, then G has even order.

Of course, there are some noteworthy exceptions to this claim about the form of statements in mathematics. For example, consider this statement:
4. There exists a function that is continuous at every real number and differentiable at no real number.

This sentence can probably be rewritten accurately in an "If-then" form. But any such rephrasing is likely to be awkward, hence most mathematicians prefer a statement similar to the one given above. Nonetheless, implicit in statement 4 are some assumptions—to be specific, the definitions of function, continuity of a function, and differentiability of a function. Statement 4 asserts that given the definitions of these concepts, a function f can be defined that is continuous everywhere and differentiable nowhere. Another example is the statement:

5. There is an infinite number of prime numbers.

Implicit in this sentence are the definitions of "prime number" and "infinite number of." Perhaps the crudest way to write this statement as an implication is as follows:

> If a prime number is a positive integer greater than 1, divisible only by 1 and itself, then, for any positive integer n, there are at least n prime numbers.

Of course this statement is somewhat awkward. A cleaner way to proceed is first to present the concept of prime number in a separate definition. Then the original statement can be phrased as: If P is the set of prime numbers, then P is infinite. The important point is that statement 5 has been written, or at least understood, as an implication.

Now to the basic topic of this unit: Given a mathematical statement in the form of an implication, "If P, then Q," how do we go about proving that the statement is true or showing that it is false?

Section 4
DIRECT PROOF

A direct proof proceeds from the hypothesis P and deduces that Q must hold. In concocting a direct proof, we are free to use all the hypotheses that are given and any statement that has been previously established. Here is an elementary example.

Example 1   Prove: The product of two odd integers is an odd integer.

First, let us translate this statement into an implication. Our initial attempts will not contain any symbols.

1. If the product of two odd integers is taken, then that product is an odd integer.

2. If two odd integers are given, then their product is an odd integer.

Now let us state a version using variables.

3. If a and b are odd integers, then $a \cdot b$ is an odd integer.

To make the quantification on a and b explicit, we have the following version:

4. For all integers a and b, if a and b are odd, then $a \cdot b$ is an odd integer.

All these statements describe the same property of the set of integers; hence none is any more "mathematical" than any of the others. While the third and fourth sentences might appear to be the most mathematical of the four, they differ from the others mainly in their conciseness and sophistication. For example, the use of "$a \cdot b$" in the conclusion must be interpreted by the reader to mean "the product of a and b." Now to the proof. By definition, an integer a is *odd* if a is not even. Using the Division Theorem (p. 66), it follows that if a is odd, then there exists an integer m such that $a = 2m + 1$.

Proof. By definition, since a and b are odd, there exist integers m and n such that $a = 2 \cdot m + 1$ and $b = 2 \cdot n + 1$. We must show that there exists an integer k such that $a \cdot b = 2 \cdot k + 1$. By familiar properties of addition and multiplication of real numbers to be described officially in Section 6,

$$a \cdot b = (2m + 1)(2n + 1)$$
$$= (2m + 1) \cdot 2n + (2m + 1) \cdot 1$$
$$= 2m \cdot 2n + 2n + 2m + 1$$
$$= 4mn + 2m + 2n + 1$$
$$= 2(2mn + m + n) + 1.$$

Thus, $a \cdot b = 2k + 1$ where k is the integer $2mn + m + n$. This proves that $a \cdot b$ is odd if a and b are odd. ∎

Remark. Here is an incorrect proof of the previous result: Let a and b be odd integers. For example, let $a = 5$ and $b = 7$. Then $a \cdot b = 35$, which is odd. Therefore, $a \cdot b$ is an odd integer. This "proof" is incorrect because it ignores the quantifications on a and b. One must show that *for all* odd integers a and b, $a \cdot b$ is odd. All this argument shows is that there exist odd integers a and b such that $a \cdot b$ is odd. To repeat what was said earlier: Remember the scope of the variables.

Next we give a direct proof of an important property of the real number system. Note again that this proof of the theorem uses properties of the real number system, such as associativity for multiplication and existence of inverses for multiplication, which are presented formally in Section 6.

Theorem 1   (Cancellation Law for Multiplication) *If a, b, and c are real numbers such that $a \neq 0$ and $ab = ac$, then $b = c$.*

Proof. Suppose a, b, and c are arbitrary real numbers such that $a \neq 0$ and $ab = ac$. Since $a \neq 0$, the multiplicative inverse of a, a^{-1}, exists. Thus
$$a^{-1} \cdot (a \cdot b) = a^{-1} \cdot (a \cdot c).$$
Hence by the associative law for multiplication,
$$(a^{-1} \cdot a) \cdot b = (a^{-1} \cdot a) \cdot c.$$
Thus
$$1 \cdot b = 1 \cdot c$$
or
$$b = c. ∎$$

The following result can be either derived from Theorem 1 or proved via a similar argument. We leave the proof as an exercise. (See Exercise 3.)

Corollary 1. *If a and b are real numbers such that $ab = 0$ and $a \neq 0$, then $b = 0$.*

Example 2   Show: If x is a real number and $x^2 - 6x + 10 = x$, then $x = 2$ or $x = 5$.

(Remark: Throughout this book we use the words "prove" and "show" interchangeably.)

Before reading on, try to prove that statement.

Proof. If $x^2 - 6x + 10 = x$, then subtracting x from both sides (i.e., adding $-x$ to both sides), we find that $x^2 - 7x + 10 = 0$. Upon factoring the left side, we see that $(x - 2) \cdot (x - 5) = 0$. Either $x - 2 = 0$ or, if not, then by Corollary 1, $x - 5 = 0$. Therefore, either $x = 2$ or $x = 5$. ∎

Conversely, it is easy to check that 2 and 5 are solutions to the given equation. Thus, for a real number x, $x^2 - 6x + 10 = x$ if and only if $x = 2$ or $x = 5$.

Our next example describes a property of sets, the so-called *distributive law of intersection over union*.

Example 3 Let A, B, and C be sets. Then

$$A \cap (B \cup C) = (A \cap B) \cup (A \cap C).$$

Proof. Using an element-chasing proof, we show that each of the given sets is a subset of the other.

To show that $A \cap (B \cup C) \subseteq (A \cap B) \cup (A \cap C)$, we let $x \in A \cap (B \cup C)$ and show that $x \in (A \cap B) \cup (A \cap C)$. Since $x \in A \cap (B \cup C)$, $x \in A$ and $x \in B \cup C$; i.e., $x \in A$ and, either $x \in B$ or $x \in C$. There are therefore two possible cases:

Case 1. $x \in A$ and $x \in B$.

Case 2. $x \in A$ and $x \in C$.

In Case 1, $x \in A \cap B$; in Case 2, $x \in A \cap C$. Since either Case 1 or Case 2 is true, either $x \in A \cap B$ or $x \in A \cap C$, and hence $x \in (A \cap B) \cup (A \cap C)$. Therefore, $A \cap (B \cup C) \subseteq (A \cap B) \cup (A \cap C)$.

Now to show that $(A \cap B) \cup (A \cap C) \subseteq A \cap (B \cup C)$, we let $x \in (A \cap B) \cup (A \cap C)$ and show that $x \in A \cap (B \cup C)$.

If $x \in (A \cap B) \cup (A \cap C)$, then $x \in A \cap B$ or $x \in A \cap C$. Consider again two cases:

Case 1. $x \in A \cap B$.

Case 2. $x \in A \cap C$.

In Case 1, $x \in A$ and $x \in B$. Since $x \in B$, either $x \in B$ or $x \in C$. Thus $x \in A$ and $x \in B \cup C$ which means that $x \in A \cap (B \cup C)$.

In Case 2, an identical argument shows that $x \in A \cap (B \cup C)$.

Thus we have proved that $x \in A \cap (B \cup C)$ whenever $x \in (A \cap B) \cup (A \cap C)$, or equivalently $(A \cap B) \cup (A \cap C) \subseteq A \cap (B \cup C)$.

Since we have inclusion in both directions, the sets are equal. ∎

Our final example describes a basic property of integers. Let a and b be integers; then a is a *factor* of b if there is an integer c such that $b = a \cdot c$. Thus, since $6 = 2 \cdot 3$ and 3 is an integer, 2 is a factor of 6. Notice that 2 is not a factor of 7 in spite of the equality $7 = 2 \cdot (7/2)$, since $7/2$ is not an integer and for each integer c, $2 \cdot c \neq 7$. Also observe that an integer n is even if and only if 2 is a factor of n. Several other ways of expressing factorability are commonly used. Instead of saying, a is a *factor* of b, one can say a *divides* b, a is a *divisor* of b, b is *divisible* by a, or b is a *multiple* of a.

Lemma 1   *Let a, b, c be integers. If a is a factor of b and of c, then a is a factor of $b + c$.*

Proof. a is a factor of b, there exists an integer n such that $b = a \cdot n$. Similarly, $c = a \cdot m$ for some integer m. Thus $b + c = a \cdot n + a \cdot m = a \cdot (n+m)$. Since $n + m$ is an integer, a is a factor of $b + c$. ∎

Notice the similarity among all these examples. In each case we reasoned directly from the hypothesis to the conclusion, using relevant definitions, facts, and rules at our disposal.

EXERCISES §4

In problems 1–7 use direct proof to establish the given statement.

1. Prove: The sum of two even integers is an even integer.
2. Prove: The product of two integers, one of which is even, is an even integer.
3. Prove Corollary 1.
4. Prove: If a and b are integers such that a is a factor of b, then, for any integer c, a is a factor of $b \cdot c$.
5. Prove: If a, b, and c are integers such that a is a factor of b and b is a factor of c, then a is a factor of c.
6. Prove: If a and b are integers and a is a factor of b, then a^2 is a factor of b^2.
7. Let a and b be integers. Prove: If a is odd and b is even, then $a + b$ is odd.

Given the right mood, one might say that indirect proof stands in direct contrast to direct proof. Remember that in almost every case, we are trying to establish an implication $P \Rightarrow Q$. In a direct proof, we start with P and deduce Q. There are two methods of indirect proof, contraposition and contradiction. While differing somewhat, these methods share the common feature of establishing the implication $P \Rightarrow Q$ by starting with not-Q. We first consider contraposition.

PROOF BY CONTRAPOSITIVE

We first discuss the *method of contraposition* or *proof by contrapositive*. This technique establishes the validity of an implication "If P then Q" by showing that its logically equivalent contrapositive "If not-Q then not-P" holds. Thus, to prove that "If P then Q" is valid, one assumes that not-Q holds and derives the conclusion that not-P is valid.

Example 1 Let n be an integer. Prove: If n^2 is an odd integer, then n is an odd integer.

Proof. To prove the statement, we prove the contrapositive: If n is an integer that is not odd, then n^2 is an integer that is not odd.

Suppose n is an integer that is not odd. Then n is even, hence $n = 2k$ for some integer k. Thus $n^2 = (2k)^2 = 4k^2 = 2(2k^2)$ is also an even integer. Therefore, n^2 is not odd if n is not odd. ∎

Observe that in this example, the proof of the contrapositive of the original statement is a direct proof. We take the meaning of the assumption that n is not odd, mix in some simple arithmetic, and conclude that n^2 is not odd. With a more complex statement, it might be necessary to use other pieces of knowledge; such complications, however, need not cloud over the basic structure of the proof by contrapositive.

Example 2 Let A and B be subsets of a set U. Show that if $A \subseteq B$, then $B^c \subseteq A^c$. (Recall that $A^c = U - A$.)

Proof. We give a proof by contraposition. We show that if $B^c \not\subseteq A^c$, then $A \not\subseteq B$. If $B^c \not\subseteq A^c$, then there exists an element $x \in B^c$ such that $x \notin A^c$. Since $x \in B^c$, it follows that $x \notin B$. Since $x \notin A^c = \{x \in U \mid x \notin A\}$, it follows that $x \in A$. Therefore x is an element of U such that $x \in A$ and $x \notin B$. Thus we conclude that $A \not\subseteq B$. ∎

In spite of the logical equivalence of the statements, "If P then Q" and "If not-Q then not-P," the application of a proof by contrapositive must be handled with care. The danger lies in the formation of negations. If P is a complex sentence, then not-P might be difficult to formulate. For example, as we have seen, the negation of a sentence involving universal and existential quantification can be especially challenging. With these words of warning in mind, we move to a related proof technique—proof by contradiction.

PROOF BY CONTRADICTION

To establish an implication "If P then Q" via the *method of contradiction*, one assumes that the statement "If P then Q" is false and attempts to derive a contradiction, namely a statement that is always false no matter what the truth values of its components. Recall that "If P then Q" is logically equivalent to "not-P or Q," its negation is therefore logically equivalent to "P and not-Q." Thus, one begins with the statement "P and not-Q," and from this statement one deduces a contradiction, usually a statement of the form "R and not-R." Thus, we show that the implication $(P \wedge \text{not-}Q) \Rightarrow (R \wedge \text{not-}R)$ is true. Since $R \wedge \text{not-}R$ is false, it follows that $P \wedge \text{not-}Q$ is also false or equivalently "not-(If P then Q)" is false and "If P then Q" is true. In most cases the statement R is either a result established in the course of the proof, a previously derived result, or an assumption made at some point in the proof.

Example 3   Show that the circle whose equation is $x^2 + y^2 = 2$ and the line with equation $y = x + 4$ do not intersect.

Proof. We rewrite the statement as an implication: If C is the circle with equation $x^2 + y^2 = 2$ and L is the line with equation $y = x + 4$, then C and L do not intersect. We give a proof by contradiction. Assume C and L do intersect. Then there exists a point (x_0, y_0) on both C and L. Then $y_0 = x_0 + 4$ and $x_0^2 + y_0^2 = 2$. Therefore, $x_0^2 + (x_0 + 4)^2 = 2$ from which it follows that $x_0^2 + 4x_0 + 7 = 0$ or that $(x_0 + 2)^2 + 3 = 0$. Thus, if (x_0, y_0) lies on the circle $x^2 + y^2 = 2$ and the line $y = x + 4$, then $(x_0 + 2)^2 = -3$. However, for any real number z, $z^2 \geq 0$, hence $z^2 \neq -3$. We have achieved a contradiction, thus confirming that the given circle and line do not intersect.   ∎

We now prove the theorem of Hippasus mentioned in Section 2.

Example 4   Prove that $\sqrt{2}$ is irrational.

Proof. We recall that a number x is rational if there exist integers m and n with $n \neq 0$ such that $x = m/n$. Therefore, to say that $\sqrt{2}$ is irrational is to say that for all integers m and n, $\sqrt{2} \neq m/n$. A direct proof that $\sqrt{2}$ is irrational would evidently require us to begin with an arbitrary pair of integers m and n and to show that $\sqrt{2} \neq m/n$. This task appears to be difficult. Thus a proof by contradiction is worth trying.

Our goal is to show that for all integers m, n with $n \neq 0$, $\sqrt{2} \neq m/n$. We assume that the negation of this statement holds, and hence we assume that there exists a pair of integers m and n for which $\sqrt{2} = m/n$. Our aim is to derive a contradiction. If m and n are both even integers, then $m = 2m'$ and $n = 2n'$ where m' and n' are integers and $\sqrt{2} = m/n = 2m'/2n' = m'/n'$. If m' and n' are even, then we repeat this procedure. In fact we continue to repeat it until we find integers a and b (which are factors of m and n respectively), *at least one of which is odd*, such that $\sqrt{2} = a/b$.

Squaring both sides and multiplying both sides of the result by b^2 gives $2b^2 = a^2$. Thus a^2 is even, hence a is even. (Why?) Therefore, $a = 2k$ and $2b^2 = a^2 = (2k)^2 = 4k^2$. By cancellation $b^2 = 2k^2$, from which it follows that b is even. We conclude that both a and b are even, a contradiction of the fact that at least one of a and b is odd. This completes the proof that $\sqrt{2}$ is irrational. ∎

Note that although the statement is not a conditional sentence, it can easily be phrased as a conditional: If x is a positive real number whose square is 2, then x is irrational.

As mentioned at the start of the argument, trying a direct proof in this case would seem to be futile. On the other hand, by assuming the negation of the desired conclusion (i.e., that $\sqrt{2}$ is rational), we obtain something quite tangible to manipulate: We are given integers m and n such that $\sqrt{2} = m/n$. In general, a proof by contradiction should be considered if negating the conclusion yields a statement that can be manipulated and exploited. For instance, in Example 1, by assuming that the given curves did intersect, we obtain a point (x_0, y_0) on both curves, and by substituting the numbers x_0 and y_0 in the equations for the curves, we derive a contradiction.

Our next example comes from the theory of numbers, the discipline of mathematics in which the behavior of whole numbers with respect to the operations of addition and multiplication is studied. An example of a statement of number theory is the two-hundred-year-old-as-yet-unproved Goldbach conjecture:

Every even number larger than two is expressible as a sum of two primes.

We illustrate the method of proof by contradiction by proving a truly ancient result in the theory of numbers, namely the previously mentioned fact that there are infinitely many primes. This result goes back to Book 7 of Euclid's *Elements*. Recall that a positive integer $n > 1$ is *prime* if the only positive integer factors of n are 1 and n. For example, the primes less than 25 are 2, 3, 5, 7, 11, 13, 17, 19, and 23. The only even positive integer that is prime is 2. Notice that if n is not prime, then n can be factored as $n = ab$ where a and b are both greater than 1. In other words, nonprimes can be written as a product of two smaller positive integers.

Once prime numbers are defined, several questions concerning primes come to mind. For instance, one can ask: Is every integer greater than 1 divisible by a prime? The next result assures us that indeed this is the case.

Lemma 1   *Let n be any integer greater than 1. Let a be the smallest factor of n that is greater than 1. Then a is prime.*

Proof. We take n and a as in the statement of the Lemma. We must show that a is prime. We give a proof by contradiction. Suppose a is not prime. Then $a = b \cdot c$ where $b > 1$ and $c > 1$. Since b is a factor of a and a is a factor of n, b is a factor of n. (See Exercise 5, Section 4, p. 41.) But b is also greater than 1 and less than a. To summarize, b is a factor of n such that $1 < b < a$. Therefore, a is not the smallest factor of $n > 1$. This contradiction (of the definition of a) means that a is indeed prime. ∎

Notice that in Example 3, the contradiction that was reached involved a property of the real numbers, whereas in Example 4 and in Lemma 1, the contradiction was of an assumed statement.

Another question one might ask about primes is the following: Suppose we begin listing the prime numbers: 2, 3, 5, 7, 11, ... Does this list ever end? Or must it continue without stopping? The next theorem answers these questions by asserting that the list of primes does not stop after a finite number of entries.

Theorem 1   *There exist infinitely many prime numbers.*

Proof. Suppose that the statement is false. Then only a finite number of primes exist. List them in increasing order: $2, 3, 5, \ldots, p$, where p is the largest prime number. Our goal is to produce a prime number that is not one of the primes on this list. Since this list allegedly contains all prime numbers, a contradiction is achieved.

To create a new prime, we exhibit a number that has no prime factor from the assumed complete list of primes, $2, 3, 5, \ldots, p$. But, by the previous lemma, this number must have at least one prime factor, call it q. The prime number q will then be a prime number that is not on the list.

We thus seek a number that has no factors among the primes $2, 3, \ldots, p$. One might be tempted to consider $p + 1$, but as examples suggest (and is easily proved), every prime factor of $p + 1$ might be on the list $2, 3, \ldots, p$. Instead, let us consider $M = (2 \cdot 3 \cdot \cdots \cdot p) + 1$.

The number M might or might not be prime, but in any case M has a prime factor by Lemma 1. Let q be a prime factor of M. (For instance, if q is the smallest factor of M greater than 1, then q is prime by the lemma.) We claim that q is not one of the primes $2, 3, \ldots, p$. For, if so, then q is a factor of $2 \cdot 3 \cdot \cdots \cdot p$ and of $M = (2 \cdot 3 \cdot \cdots \cdot p) + 1$. It follows from the lemma in Section 4 that q is a factor of $M - (2 \cdot 3 \cdot \cdots \cdot p) = 1$. But no integer greater than one can be a factor of one. Thus, q is not one of the primes $2, 3, \ldots, p$. (Note that we have just given a proof by contradiction within a proof by contradiction.)

Therefore, we have produced a prime number that is different from each of the primes $2, 3, \ldots, p$. But this conclusion contradicts the assumption that the list $2, 3, \ldots, p$ contains every prime number.   ∎

Technical Remark. Instead of taking $M = (2 \cdot 3 \cdot \cdots \cdot p) + 1$, we could have chosen $M = p! + 1 = (p \cdot (p - 1) \cdot (p - 2) \cdot \cdots \cdot 3 \cdot 2 \cdot 1) + 1$. The important point is that no integer larger than 1 can divide two consecutive integers. Thus none of the primes 2, 3, $\ldots$, p divides $p! + 1$, and hence any prime factor of $p! + 1$ is different from 2, 3, $\ldots$, p.

Cultural remark. One might ask if a quantitative version of Theorem 1 exists. For instance, for each positive integer n, can we give a formula for the number of primes $\leq n$? Following custom, we let $\pi(n)$ denote the number of primes $\leq n$. Thus we are asking if there is a nice formula for the function $\pi(n)$. This question was first considered by the great German mathematician C. F. Gauss in 1791 when he was 14 years old. Gauss did not give a precise formula for $\pi(n)$ but he did discover the nature of the long-term growth of $\pi(n)$. By inspecting tables of primes, Gauss conjectured that $\pi(n)$ is approximately $n/\log(n)$, where $\log(n)$ is the natural logarithm of n. Specifically, Gauss conjectured that

$$\lim_{n \to \infty} \pi(n)/[n/\log(n)] = 1.$$

This intriguing result, known as the Prime Number Theorem, was first proved in 1896 by two French mathematicians, Hadamard (whom we met at the start of this section) and de la Valleé Poussin. Note that one corollary of the Prime Number Theorem is that $\lim_{n \to \infty} \pi(n) = \infty$, and hence the number of primes is infinite.

Concluding remarks. The two methods of indirect proof, contradiction and contraposition, possess notable similarities and differences. In both methods the negation of the conclusion of the implication "If P then Q," namely not-Q, is assumed to hold. In a proof by contrapositive, the goal is to derive the statement not-P from the statement not-Q. In a proof by contradiction, we attempt to extract some contradiction from the statement "P and not-Q." In a proof by contradiction, we have more to work with— both P and not-Q—than we have in a proof by contrapositive—only not-Q. But we pay a price for the extra weapon. In a proof by contrapositive, we have the advantage of aiming for a specific goal, not-P. By contrast, with the contradiction method, we shoot only for some contradiction. Very often we do not know what shape this contradiction will take when we start the proof. Perhaps it is this fact that can make a proof by contradiction difficult to understand (upon reading) and to create.

As is always the case in mathematics and science, a technique, once mastered, must be applied judiciously. A common question is: When should a particular proof technique be applied? For better or worse, there seems to be no definitive answer to this question, but as far as the contradiction and contrapositive methods are concerned, there are some useful guidelines to observe.

(1) Consider a contradiction or contrapositive argument if the conclusion is difficult to grasp or manipulate as is, and more information and material to play with are obtained by negating the conclusion.

This advice was followed with success in all the examples presented. For example, in the proof of Theorem 1, the negation of the existence of infinitely many primes enables us to manipulate the finite set of all primes. We can then use addition and multiplication along with properties of these operations to derive a contradiction.

(2) Consider a contradiction or contrapositive argument if the conclusion is the negation of some other statement.

Thus if the conclusion has the form not-$P(x)$, then one would assume that $P(x)$ holds and would derive a contradiction. For example, to show that $\sqrt{2}$ is not rational, we assume that $\sqrt{2}$ is rational and derive a contradiction. In Example 1, we assume that the given circle and line intersect and derive the contradiction that there exists a real number z such that $z^2 = -3$.

(3) Consider a contradiction or contrapositive argument if the conclusion demands either a unique object or at most one object having a given property.

For a problem asking for a unique object that has a given property, one usually breaks the argument down into two steps:

  (i) Show that there is at least one object having the property;
  (ii) Show that there is at most one object having the property.

In establishing these steps, especially step (ii), contradiction or contraposition can be useful. The advice given in guideline 3 is perhaps implicit in guideline 1. Nevertheless, since statements asserting the uniqueness of an object are common in mathematics, it is worth noticing that such statements tend to be susceptible to either contraposition or contradiction.

EXERCISES §5

1. Let a be an integer. Prove: If a^2 is even, then a is even.
2. Let a and b be integers. Show: If $a \cdot b$ is odd, then a is odd and b is odd.
3. Prove: $\sqrt{8}$ is irrational.
4. (a) Prove: $\sqrt{6}$ is irrational.
   (b) Prove: $\sqrt{10}$ is irrational.
5. Prove: $\log_2(3)$ is irrational. (Recall that $\log_2(3) = x$ means that $2^x = 3$.)
6. Prove: $\log_2(5)$ is irrational.
7. Prove: The product of a nonzero rational number and an irrational number is an irrational number.
8. Prove: If x is irrational, then $\sqrt{x}$ is irrational.
9. Prove: If $A \subseteq B$ and $B \subset C$, then $A \subset C$.

Section 6

AN EXAMPLE:
THE REAL NUMBERS

At this point in our discussion, we pause to consider again the set of real numbers. Our ostensible goal is to develop the general notion of an axiomatic system, a concept that has become fundamental in mathematics in the past century, and to give an axiomatic presentation of the real numbers, perhaps the most important example of an axiomatic system. Simultaneously, we kill another bird by using the real numbers to illustrate the proof techniques thus far discussed in this text.

The first axiom system to appear in the history of mathematics was Euclid's development of plane geometry. Euclid's treatment begins with some *primitive* or *undefined* terms. Next, a list of statements that are assumed to be true is added. Some of these statements are called *axioms*, some are called *postulates*. The collection of undefined terms, axioms, and postulates forms the *axiom system* for Euclidean geometry. Let us look at Euclid's system more closely.

The undefined terms in Euclidean geometry are *points* and *lines*. The *axioms*, or self-evident truths as they were sometimes called, are general statements whose truth cannot presumably be disputed. Examples are

The whole is greater than the part,

and

Equals added to equals yield equals.

The *postulates* are statements that describe relationships among the undefined terms of the system. Hence, the scope of a postulate is more limited than that of an axiom. Examples of postulates are

Two points determine exactly one line,

and

Through a point not on a given line, one and only one line can be drawn that does not meet the given line.

From this collection of undefined terms, axioms, and postulates, one can deduce through logical reasoning a large number of statements (called *theorems, propositions*, or *lemmas*) about plane geometry, a notable example being: The sum of the angles of a triangle is π radians where π is the ratio of the circumference of any circle to the diameter of that circle. In summary, the Euclidean axiom system consists of a collection of undefined terms, axioms, and postulates. The body of statements that can be deduced from the axioms and postulates constitutes the Euclidean theory of plane geometry.

Until the middle of the nineteenth century, Euclidean geometry held a unique place in mathematics. First, Euclidean geometry was regarded by most learned people as the true geometry of the universe. Its axioms and postulates were indeed basic truths of the physical world. Thus all the theorems of Euclidean geometry were also true descriptions of physical reality. Second, until this time no other portion of mathematics was treated axiomatically. Whenever possible, mathematicians attempted to deduce results in other areas of mathematics (such as calculus) from Euclidean geometry. Otherwise, they used some seemingly obvious results (such as $x^2 - y^2 = (x - y)(x + y)$, where x and y are real numbers) without a careful regard for the status of such statements: Is $x^2 - y^2 = (x - y)(x + y)$ a postulate of the real number system, does it follow from some other postulates and axioms for the real numbers, or is it an empirical fact about real numbers?

During the nineteenth century, two important developments occurred. First, for reasons both internal and external to mathematics, mathematicians began to question the status of Euclidean geometry. Are the postulates and theorems actually truths of nature, or are the theorems just statements deducible from the axioms and postulates that are in turn merely assumptions made by mathematicians? These questions led a few mathematicians, notably Lobatchevsky, Bolyai, and Gauss, to modify the postulates of Euclidean geometry to obtain other geometric theories. While these non-Euclidean geometries produced results that seemed counterintuitive, they did appear to be logically consistent. Thus the existence of axiom systems in mathematics other than Euclidean geometry became a mathematical reality.

The second major development in the nineteenth century was the rapid expansion of several other areas of mathematics. Due to pressures both inside and outside mathematics, disciplines such as number theory, algebra, and analysis grew significantly in extent and sophistication. Mathematicians felt the need to clarify and to organize the results of each of these complex subjects. To do so, mathematicians such as H. Grassmann, R. Dedekind, K. Weierstrass, and G. Peano turned to the axiomatic method. Their goal was to develop each mathematical theory from a collection of primitive terms, axioms, and postulates. Initially, these mathematicians were interested primarily in clearing up the development of previously existing bodies of mathematical knowledge. As mathematicians became more adept at using axiomatic systems, they came to use them as more than merely organizing devices. In fact, during the twentieth century, many mathematical theories have from their inception been developed on an axiomatic basis. Thus, the axiomatic method has become a mathematical research tool.

The modern view of axiomatic systems is remarkably close to the Euclidean conception of axiomatic geometry. One important change is that the distinction between axioms and postulates has disappeared in the past 100 years. This conceptual shift has been yet another result of the advances made in mathematics in the past 100 years. Many of Euclid's "self-evident" truths have been seen to be not at all evident. For example, there are situations in which a whole is not necessarily greater than one of its parts (see Chapter 4 of [2]). According to the modern view, then, an *axiom system* consists of primitive or undefined terms together with axioms, which are statements describing relationships among and properties of the primitive concepts. (To emphasize the arbitrariness of the undefined terms, the influential mathematician David Hilbert once asserted: "One must be able to say at all times—instead of points, straight lines, and planes—tables, chairs, and beer mugs.") The axiom system and all the theorems deducible from the axioms together form the *axiomatic theory*.

An axiomatic theory can be likened to the everyday concept of a game. Consider any familiar game, such as chess or checkers. The pieces, the gameboard, and the players of the game constitute the primitive terms. The rules of each game describe how the pieces interact; thus, the rules play the role that the axioms play in an axiomatic theory. All the possible legal maneuvers and configurations in each game are derivable from the rules; hence, these maneuvers and configurations are analogous to the theorems of an axiomatic theory.

A note about the names of statements that are deducible from the axioms. The most important of these statements is called a *theorem*. A statement that is derivable from axioms and considered less significant or perhaps more obvious than theorems is often called a *proposition* (not to be confused with the propositions of propositional calculus); a statement that is technical in nature and of a limited scope is called a *lemma*. Finally, a statement that follows immediately from a theorem, in some cases being merely a special case of a theorem, is called a *corollary*. From a logical point of view, theorems, lemmas, propositions, and corollaries are on an equal footing. Nonetheless, the labels, while admittedly being somewhat arbitrary, enable us to clarify their position in the development of a particular subject.

We now give an axiomatic description of the real number system. The axioms for **R** fall into three classes according to the types of properties that they describe: field properties, order properties, and the completeness property.

The *real numbers* consist of a collection or set, **R**, of objects called *numbers* satisfying the following fourteen conditions.

FIELD AXIOMS

1. (Closure for addition)   For each pair $x, y \in \mathbf{R}$, there exists a unique object in **R**, written $x + y$ and called the **sum** of x and y.

2. (Associative law for addition)   For all $x, y, z \in \mathbf{R}$, $(x + y) + z = x + (y + z)$.

3. (Additive identity)   There is an object $0 \in \mathbf{R}$ such that for all $x \in \mathbf{R}$, $x + 0 = 0 + x = x$.

4. (Additive inverse)   For each $x \in \mathbf{R}$, there exists $y \in \mathbf{R}$ such that $x + y = y + x = 0$.

5. (Commutative law for addition)   For all $x, y \in \mathbf{R}$, $x + y = y + x$.

6. (Closure for multiplication)   For all $x, y \in \mathbf{R}$, there exists a unique object in $\mathbf{R}$, written $x \cdot y$ and called the *product* of x and y.

7. (Associative law for multiplication)   For all x, y, $z \in \mathbf{R}$, $(x \cdot y) \cdot z = x \cdot (y \cdot z)$.

8. (Multiplicative identity)   There exists an object $1 \in \mathbf{R}$ such that $1 \neq 0$ and for all $x \in \mathbf{R}$, $x \cdot 1 = 1 \cdot x = x$.

9. (Multiplicative inverse)   For each $x \in \mathbf{R}$ such that $x \neq 0$, there exists $y \in \mathbf{R}$ such that $x \cdot y = y \cdot x = 1$.

10. (Commutative law for multiplication)   For all $x, y \in \mathbf{R}$, $x \cdot y = y \cdot x$.

11. (Distributive law for multiplication over addition)   For all $x, y, z \in \mathbf{R}$, $x \cdot (y + z) = x \cdot y + x \cdot z$.

Remark on inverses: Axiom 4 asserts that for each $x \in \mathbf{R}$, there exists at least one $y \in \mathbf{R}$ such that $x + y = y + x = 0$. One can go a step further by showing that for any $x \in \mathbf{R}$ there exists at most one y such that $x + y = y + x = 0$. (See Theorem 1.) This real number y is called *the additive inverse of* x, denoted by $-x$. In addition, we define $x - y$ to be $x + (-y)$. Similarly one can show that each $x \in \mathbf{R}$ such that $x \neq 0$ has a unique *multiplicative inverse*. This element is denoted by x^{-1}. For $y \neq 0$ we define x/y to be $x \cdot y^{-1}$.

ORDER AXIOMS

There exists a subcollection $\mathbf{R}^{+}$ of $\mathbf{R}$ called the *positive numbers* satisfying the following conditions:

12. If x and y are in $\mathbf{R}^{+}$, then $x + y$ and $x \cdot y$ are also in $\mathbf{R}^{+}$.

13. For each $x \in \mathbf{R}$, exactly one of the following three conditions holds: (i) $x \in \mathbf{R}^{+}$, (ii) $x = 0$, or (iii) $-x \in \mathbf{R}^{+}$.

If $x \in \mathbf{R}$ and $-x \in \mathbf{R}^{+}$, then x is called a *negative number*.

Axioms 12 and 13 allow us to propose the following definition, which describes how the members of $\mathbf{R}$ can be compared or ordered.

Definition 1   *Ordering of* **R**

Let x and y be in **R**. We say that x *is greater than* y, written $x > y$, if $x - y = x + (-y)$ is in $\mathbf{R}^+$. We say that x *is greater than or equal to* y, written $x \geq y$, if $x > y$ or $x = y$. We say that x *is less than* (respectively *less than or equal to*) y, written $x < y$ (respectively $x \leq y$), if $y > x$ (respectively $y \geq x$).

This concept of ordering in turn allows us to introduce the notion of boundedness, that is needed for the statement of the completeness property.

Definition 2   *Boundedness*

A collection A of real numbers is called *bounded from above* if there exists a real number x (not necessarily in A) such that for all a in A, $x \geq a$. Such an element x is called an *upper bound* for A.

Note the order of the quantification. A collection A is bounded from above if there exists at least one number x that is greater than or equal to all the members of A.

Definition 3   *Least upper bound*

Let A be a set of real numbers that has an upper bound. An element x_0 of **R** is called a *least upper bound* of A if (i) x_0 is an upper bound for A, and (ii) for any upper bound x of A, $x_0 \leq x$.

As an example, consider the set A of all real numbers that are less than 2. A is bounded from above by 2, 3, or 4, for instance. Observe that 2 is a least upper bound for A. With these concepts in hand, we can state the final axiom for **R**.

COMPLETENESS AXIOM

14. Any *nonempty* set of real numbers having an upper bound has a least upper bound.

At this point let us summarize formally our definition of the real numbers.

Definition 4   *The real numbers*
The *real numbers* is a set of objects, denoted by **R**, which satisfies Axioms 1–14.

This definition suggests an important question: Is there actually a collection of objects that satisfies Axioms 1–14? It is at least conceivable that the axioms are somehow self-contradictory, and hence can be satisfied by no set of mathematical objects. If we picture the real numbers as points along a line or if we represent real numbers in decimal form, then perhaps we can be convinced that the real numbers do satisfy all the axioms given above. Such an argument can be suggestive but is hardly convincing. To demonstrate mathematically that the real numbers, **R**, do exist, we must

 (i) define the set of objects that constitute **R**,
 (ii) define, for all x and y in **R**, the sum $x + y$,
 (iii) define, for all x and y in **R**, the product $x \cdot y$,
 (iv) define the objects that constitute the collection $\mathbf{R}^+$ of positive numbers,
 (v) show that Axioms 1–14 hold for **R** with addition, multiplication, and positivity as defined in (i)–(iv).

In [2] it is shown how a set that satisfies Axioms 1–14 can actually be defined. In the language of mathematical logic, this result shows that the axiom system consisting of Axioms 1–14 has a *model*. Moreover, it can be proved that in a sense only one collection satisfies these axioms: If two collections of objects $\mathbf{R}_1$ and $\mathbf{R}_2$ are defined, and if $\mathbf{R}_1$ and $\mathbf{R}_2$ both satisfy Axioms 1–14, then the objects in $\mathbf{R}_1$ and $\mathbf{R}_2$ can be matched exactly in such a way that addition, multiplication, and positivity are preserved. In this case we say that $\mathbf{R}_1$ and $\mathbf{R}_2$ are *isomorphic*. Thus, $\mathbf{R}_1$ and $\mathbf{R}_2$ are identical in terms of their properties although the labelings of their elements may differ. In technical terms this result shows that Axioms 1–14 have a unique model. Therefore, we can legitimately call any collection satisfying the fourteen postulates **the real numbers**.

For now let us derive some elementary results from the axioms. Our first theorem asserts that each element has at most one additive inverse. This result and Axiom 4 imply that each element of **R** has a unique additive inverse.

Theorem 1   *Let x be any real number. Suppose y_1 and y_2 are elements of* **R** *that satisfy the equations $x + y_1 = y_1 + x = x + y_2 = y_2 + x = 0$. Then $y_1 = y_2$.*

Proof. By assumption $x + y_1 = 0$. Therefore $y_2 + (x + y_1) = y_2 + 0 = y_2$ by Axiom 3. By Axiom 2, $y_2 + (x + y_1) = (y_2 + x) + y_1$. Thus $y_2 = (y_2 + x) + y_1 = 0 + y_1 = y_1$ by our assumption on y_2 and by Axiom 3. ∎

As noted earlier, because for each $x \in \mathbf{R}$ there exists exactly one y in $\mathbf{R}$ such that $x + y = y + x = 0$, we can denote y by $-x$ without fear of ambiguity: $-x$ is the unique element of $\mathbf{R}$ whose sum with x is zero. We also observe that this proof uses only Axioms 1–3, hence our use of $-x$ in later axioms is legitimate. The analogous fact for multiplication can be derived in a similar way: If $x \neq 0$, then there exists exactly one y in $\mathbf{R}$ such that $x \cdot y = y \cdot x = 1$. This element y is denoted by x^{-1}.

Next we derive some order properties of $\mathbf{R}$.

Theorem 2 *(i) Let x and y be in $\mathbf{R}$. Then exactly one of the following conditions holds: (a) $x > y$, (b) $x = y$, or (c) $x < y$.*

(ii) Let x, y, and z be in $\mathbf{R}$. If $x < y$ and $y < z$, then $x < z$.

Proof. (i) Consider $x - y$. By Axiom 13, exactly one of the following conditions holds: (a) $x - y$ is in $\mathbf{R}^+$, (b) $x - y = 0$, or (c) $-(x - y)$ is in $\mathbf{R}^+$.

If $x - y$ is in $\mathbf{R}^+$, then by definition $x > y$. If $x - y = 0$, then $(x - y) + y = 0 + y = y$ and, therefore, $y = (x - y) + y = (x + (-y)) + y = x + ((-y) + y) = x + 0 = x$. Finally note that $-(x - y) = y - x$ (see Exercise 4), hence if $-(x - y) = y - x$ is in $\mathbf{R}^+$, then $x < y$.

(ii) If $x < y$ and $y < z$, then $y - x$ and $z - y$ are in $\mathbf{R}^+$. By Axioms 2–4, $(z - y) + (y - x) = ((z - y) + y) + (-x) = (z + (-y + y)) + (-x) = (z + 0) + (-x) = z + (-x) = z - x$. By Axiom 12, $z - x = (z - y) + (y - x)$ is in $\mathbf{R}^+$, and therefore $x < z$. ∎

The real number system is, of course, a very familiar object. After having worked with this system for so many years, it might seem strange for us to be proving "obvious" facts about it. At the same time it might be unclear exactly what can be assumed about $\mathbf{R}$ and what must be proved about $\mathbf{R}$. For example, how do we know that if a, b, and c are real numbers and $a = b$, then $a + c = b + c$? Is this a property that one merely assumes? Or does it follow from the axioms? The answer is that it follows from the axioms. For the pair of numbers a and c and the pair b and c are identical. Therefore by Axiom 1, the number $a + c$ must coincide with $b + c$.

In general, any of the familiar properties of the real numbers can be deduced from Axioms 1–14.

We now consider the completeness property. This property is clearly the most complex of all given in the axioms. For starters, its statement is technical in nature, requiring the concepts of upper bound and least upper bound. Nonetheless the completeness property can be visualized rather easily as we shall see at the end of this section. For now let us deduce from the completeness axiom an important property of the real numbers, namely the fact that the square root of 2 exists in $\mathbf{R}$. (For the record, the element 2 in $\mathbf{R}$ is defined to be $1 + 1$ where 1 is the multiplicative identity in $\mathbf{R}$.)

Theorem 3 *There exists a real number u such that $u^2 = 2$.*

The idea behind the following proof is to define a set of real numbers A such that A is bounded and the least upper bound of A (which exists by axiom 14) is a number whose square is 2. The proof is rather subtle and must be read carefully. At least twice in the course of the argument, a proof by contradiction is used.

Proof. Let A be the collection of all real numbers whose square is less than 2. Thus x is in A if and only if $x^2 < 2$. First note that A is nonempty since $0 \in A$. Next observe that the collection A has an upper bound: For example, if x is in A, then $x < 2$. (Using the axioms for $\mathbf{R}$, one can show that if $x \geq 2$, then $x^2 \geq 4$, hence x is not in A.) Hence 2 is an upper bound for A. A similar argument shows that 1.5 and 1.42 are also upper bounds for A.

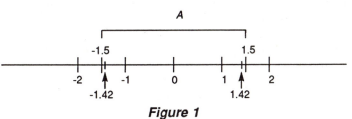

Figure 1

By Axiom 14, A has a least upper bound, call it u. We claim that $u^2 = 2$.

By Theorem 2, exactly one of the following possibilities holds: (i) $u^2 = 2$, (ii) $u^2 < 2$, or (iii) $u^2 > 2$. We show that each of the last two possibilities leads to a contradiction. For future reference we note that $1.41 < u < 1.42$.

First we suppose $u^2 < 2$. We show that contrary to our assumption, u is not an upper bound for A by finding a number that is in A and that is slightly larger than u.

Let $a = 2 - u^2$. Since $u > 1.41$, $0 < a = 2 - u^2 < 2 - (1.41)^2 < .02$. Now consider the real number $u + a/6$. We show that $u + a/6$ is in A and $u < u + a/6$. This conclusion will imply that u is not an upper bound for A, which is contrary to the definition of u. Since this invalid conclusion is implied by the assumption that $u^2 < 2$, this assumption cannot be true.

We show $u + a/6$ is in A by proving that $(u + a/6)^2 < 2$. By Exercises 5 and 6,

$$(u + a/6)^2 = u^2 + ua/3 + a^2/36.$$

Our strategy is to estimate the numbers $ua/3$ and $a^2/36$ in order to show that $(u + a/6)^2 < 2$. Since $u < 1.42$, $u/3 < 1/2$, and hence $ua/3 < a/2$. Also $a/36 < .02/36 < 1/2$ and $a^2/36 < a/2$. Thus

$$(u + a/6)^2 = u^2 + ua/3 + a^2/36 < u^2 + a/2 + a/2 = u^2 + a = 2.$$

Therefore, $u+a/6$ is in A. Finally, since $a > 0$, $a/6 > 0$ and $u+a/6 > u$. We now know that $u^2 \not< 2$, since the assumption that $u^2 < 2$ led to a contradiction.

Suppose next that $u^2 > 2$. Let $b = u^2 - 2$. Since $u < 1.42$, $0 < b < (1.42)^2 - 2 < .02$. Consider the real number $u - b/6$. We show that $(u - b/6)^2 > 2$, which implies that $u - b/6$ is an upper bound for A; but since $u - b/6 < u$, it follows that u is not a least upper bound for A.

We have

$$(u - b/6)^2 = u^2 - ub/3 + b^2/36 > u^2 - ub/3 > u^2 - 2b/3 > u^2 - b = 2$$

Thus $u-b/6$ is an upper bound for A that is less than u. Therefore, $u^2 \not> 2$, since the assumption that $u^2 > 2$ led to a contradiction.

As a result we conclude $u^2 = 2$. ∎

Theorems 1–3 illustrate the nature of axiomatic systems. Each theorem is deduced from certain of the axioms of the real number system. For instance in proving Theorem 3, we did not explicitly describe a member of **R** whose square is 2. Instead we showed how the existence of that element follows from the axioms of **R** and the rules of logic. Thus once we know that a collection of objects satisfying Axioms 1–14 exists, then we know that in that collection an element whose square is 2 exists.

We remarked earlier that **R** is in a sense the only set that satisfies Axioms 1–14. In the technical language of the foundations of mathematics, this claim asserts that the axiom system given by Axioms 1–14 has a unique *model.* If we consider only the field axioms, namely Axioms 1–11, then we can find many models that are essentially different. (The technical word is *non-isomorphic.*) More specifically, there are sets of objects, F_1 and F_2, which satisfy Axioms 1–11 with the property that the elements of F_1 and F_2 cannot be matched in such a way that addition and multiplication are preserved by the matching. Any set satisfying Axioms 1–11 is called a *field.* For now let us give one other example of a field, namely the collection of rational numbers **Q**.

By definition a real number x is rational if $x = m/n$ where m and n are integers and $n \neq 0$. Rational numbers are added and multiplied according to the following well-known formulas:

$$m/n + m'/n' = (mn' + m'n)/nn'$$

$$(m/n) \cdot (m'/n') = mm'/nn'.$$

From these definitions it follows that the sum of two rational numbers is rational and the product of two rational numbers is rational. One says that **Q** is *closed* under addition and multiplication. (For example, the collection of all rational numbers with denominators at most two, when put in lowest terms, is closed under addition but is not closed under multiplication.) Also the sum (product) of m/n and m'/n' as defined above coincides with the sum (product) of m/n and m'/n' within the real number system. Thus addition and multiplication of rational numbers satisfies the associative, commutative, and distributive properties. Moreover, the real numbers $0 = 0/1$ and $1 = 1/1$ are also rational. Finally, for any x in $\mathbf{Q}$, $x = m/n$, the additive inverse of x, $-x = -m/n$, is also in $\mathbf{Q}$ and for $x = m/n \neq 0$, the multiplicative inverse of x, $x^{-1} = n/m$, is also in $\mathbf{Q}$. Therefore, the collection $\mathbf{Q}$ of rational numbers satisfies Axioms 1–11. In other words, Axioms 1–11 are valid if $\mathbf{R}$ is replaced by $\mathbf{Q}$ throughout and addition and multiplication are defined as above. The collection of rational numbers then provides another example of a mathematical object that satisfies the field axioms. In other words, $\mathbf{Q}$ is a model for the axiom system consisting of Axioms 1–11.

Also notice that Axioms 12 and 13 hold when $\mathbf{R}$ is replaced by $\mathbf{Q}$. As a result $\mathbf{Q}$ is called an *ordered field*. At this point our luck runs out. $\mathbf{Q}$ does not satisfy Axiom 14. Our argument is based on the fact that $\sqrt{2}$ is not in $\mathbf{Q}$, which was proved in Section 5. Therefore, the set A of all rationals such that $x^2 < 2$ has an upper bound in $\mathbf{Q}$ (for example, 2 is an upper bound for A) but does not have a least upper bound in $\mathbf{Q}$. (Remember $\sqrt{2}$ is a least upper bound for A in $\mathbf{R}$.) To sum up, while $\mathbf{Q}$ is an ordered field, $\mathbf{Q}$ is not a complete ordered field.

We close this section by connecting the informal and formal developments for the real numbers given thus far. We give a special emphasis to geometric interpretations of the axioms for $\mathbf{R}$.

Logically speaking, it is unnecessary for us to bother with geometric interpretations of the axioms: We need only our deductive faculties and not our intuition in order to prove theorems about the real numbers. What then is the role of the concrete, informal representations of $\mathbf{R}$ presented earlier in this section? First, the geometric and decimal representations of real numbers provide a working grasp of $\mathbf{R}$ that is adequate for most high school and college mathematics courses. More importantly, the intuitive ways of looking at $\mathbf{R}$ are helpful (if not necessary) when we seek to *discover* new properties of $\mathbf{R}$. At times, more insight is gained from a crude but general picture of a mathematical entity than from a precise but narrow description of it. In any case, by combining the informal representations with the formal axiomatic development, we gain a deeper understanding of the real number system.

First we interpret positivity geometrically. The positive real numbers correspond to those points on the number line lying to the right of the point 0. In general, $x < y$ means that the point x lies to the left of y on the number line.

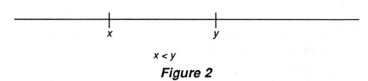

$$x < y$$

Figure 2

Next, if x and y are positive real numbers, then we can interpret $x + y$ as follows: Take a copy of the line segment from 0 to x and move the segment so that its left end is at y. The right end point of the segment is at the point $x + y$.

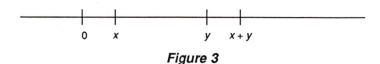

Figure 3

Now what about the concepts of upper bound and least upper bound? Recall that an upper bound z for a set of numbers A has the property that $z \geq x$ for all x in A. Thus z lies to the right of A. The least upper bound u of A is essentially the *first* real number to the right of A. This statement, however, requires some clarification. Since u is the least upper bound of A, no real number less than u is an upper bound for A: If $y < u$, then y is not an upper bound for A; thus there exists an x in A such that $y < x \leq u$. If u is in A, then for any $y < u$, we can always take $x = u$. Thus if u is in A, then u is the right end of A. If u is not in A, then u lies to the right of A and there are points of A arbitrarily close to u just to the left of u (Figure 4).

No elements of A are $> u$

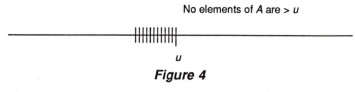

$$u$$

Figure 4

Our axiomatic presentation of the real number system is now complete.

EXERCISES §6

1. Prove that if $x \in \mathbf{R}$ and $x \neq 0$, then there exists a unique element $y \in \mathbf{R}$ such that $x \cdot y = y \cdot x = 1$.
2. (a) Prove: For all $x \in \mathbf{R}$, $x \cdot 0 = 0$.
   (b) Prove: For all $x \in \mathbf{R}$, $-(-x) = x$.
   (c) Prove: For all $x, y \in \mathbf{R}$, $x \cdot (-y) = (-x) \cdot y = -(x \cdot y)$.
   (d) Prove: For all $x \in \mathbf{R}$, $(-1) \cdot x = -x$.
3. (a) Prove: For all $x \in \mathbf{R}$, $(-1) \cdot (-x) = x$.
   (b) Prove: $(-1) \cdot (-1) = 1$.
   (c) Prove: For all $x \in \mathbf{R}$, $(-x) \cdot (-x) = x \cdot x$.
4. Prove: For all $x, y \in \mathbf{R}$, $-(x - y) = y - x$.

5. Prove: For all $x, y \in \mathbf{R}$, $(x + y)^2 = x^2 + 2 \cdot x \cdot y + y^2$. (Just for the record, $a^2 = a \cdot a$ and $2 \cdot a = (1 + 1) \cdot a = a + a$.)

6. (a) Prove: For $x, y \in \mathbf{R}$ if $x < y$ then $-x > -y$.
   (b) Prove: For x, y, z, $w \in \mathbf{R}$ if $x < y$ and $z < w$, then $x + z < y + w$.

7. Let $\mathbf{R}^+$ be the collection of positive reals.
   (a) Using Axioms 12 and 13, show that $1 \in \mathbf{R}^+$ and $-1 \notin \mathbf{R}^+$.
   (b) Show that if $x \in \mathbf{R}^+$, then $x^{-1} \in \mathbf{R}^+$.

8. Propose a definition for each of the following concepts.
   (a) A collection A of numbers is *bounded from below*.
   (b) A *lower bound* for a set of numbers A.
   (c) A *greatest lower bound* for a collection of numbers A.
   (d) Give an example of a collection A of real numbers that has a lower bound. Does this collection have a greatest lower bound?

9. (a) Let A be any collection of real numbers having a lower bound. Let B consist of all those real x such that $-x$ is in A. Prove that B has an upper bound.
   (b) Using the axioms of $\mathbf{R}$, prove that any nonempty set A of real numbers having a lower bound has a greatest lower bound.

10. Show that if x_0 and x_1 are both least upper bounds for a set A of real numbers, then $x_0 = x_1$.

11. (a) Show that 5 is the least upper bound of the set of all real numbers that are less than 5.
    (b) Let A be the set of real numbers x such that $0 < x < 2$ and $x \neq 1$. Show that A does not satisfy the completeness axiom. In other words, show that there is a collection of elements of A that is bounded above in A but that does not have a least upper bound in A.

12. Consider the following axiomatic system. The primitive or undefined terms are *person* and *collection*. On the basis of these terms, the following concepts are defined:
    (i) A *committee* is a collection of one or more persons.
    (ii) A person in a committee is called a *member* of that committee.
    (iii) Two committees are equal if and only if every member of the first committee is a member of the second and vice versa.
    (iv) Two committees having no members in common are called *disjoint* committees.

    The axioms of the system are
    (i) Every person is a member of at least one committee.
    (ii) For every pair of distinct persons there is one and only one committee of which both are members.
    (iii) For every committee there is one and only one committee that is disjoint from it.

    (a) Prove: Every person is a member of at least two committees.

    (b) Prove: There are at least four persons in the entire collection, provided there is at least one person in the collection.

13. Prove: If a, b, and c are real numbers such that $a > 1$, $b > 1$, and $ab = c$, then $a < c$ and $b < c$.

14. Let a, b, c, and d be real numbers. Prove:
    (a) if $a > b$ and $c > 0$, then $ac > bc$.
    (b) if $a > b > 0$ and $c > d > 0$, then $ac > bd$.

Section 7
MATHEMATICAL INDUCTION

Our next method of proof, the principle of mathematical induction, arises in a variety of situations throughout mathematics and computer science. Often, in the course of analyzing a mathematical phenomenon, one encounters statements of the following form:

1. For every positive integer n, $1 + 2 + \cdots + n = n(n + 1)/2$.
2. For any positive integer n, $1 + x + \cdots + x^n = (1 - x^{n+1})/(1 - x)$ for any real number x not equal to 1.

Each of these statements entails an assertion about every positive integer. For example, when $n = 5$, statement 1 says that $1+2+\cdots+5 = 5 \cdot 6/2$, which happens to be true. When $n = 2$, statement 2 reads: $1 + x + x^2 = (1 - x^3)/(1 - x)$ for each real number $x \neq 1$. This equality does indeed hold for all $x \neq 1$.

Returning to statement 1, we ask: Is it true that for each positive integer n, $1 + 2 + \cdots + n = n(n + 1)/2$? To investigate the matter further, we can calculate $1 + 2 + \cdots + n$ for $n = 1, 2, \ldots, 10$. Doing so, we obtain the sequence $1, 3, 6, 10, 15, 21, 28, 36, 45, 55$. If we calculate $n(n + 1)/2$ for $n = 1, 2, \ldots, 10$, then we obtain the identical list of ten integers. By now we probably believe that statement 1 is true for every positive integer n. To confirm our belief a bit more, we can check more cases such as $n = 11$ and $n = 12$ and we find that statement 1 is also true in these instances. Thus for any positive integer $n \leq 12$, $1 + 2 + \cdots + n = n(n + 1)/2$. But how can we show that this equality holds *for every positive integer n*? To prove statements 1 and 2, we appeal to a property of the set of positive integers known as the *Principle of Mathematical Induction* (PMI). We state PMI in a general form and show how it applies to statements 1 and 2. Later we consider a slightly different (but logically equivalent) version of mathematical induction called the Principle of Strong Induction.

Principle of Mathematical Induction

Let $S(1), S(2), \ldots, S(n), \ldots$ be a list of statements, one for each positive integer. Then every statement on the list is true if the following two conditions hold:

(i) $S(1)$ is a true statement,

(ii) For each positive integer k, if $S(k)$ is true, then $S(k + 1)$ is true.

PMI can be regarded as a proof technique that is applicable whenever one is attempting to prove that each statement in an infinite list of statements is true. Why are conditions (i) and (ii) sufficient to imply that $S(n)$ is true for all positive integers n? The first statement merely says that statement $S(1)$ is true. The second asserts that if any particular statement is true, then the next statement on the list is also true. Therefore, knowing by (i) that $S(1)$ is true and using (ii), one can conclude that $S(2)$ is true. From (ii) it follows that since $S(2)$ is true, $S(3)$ is true. By another application of (ii), one sees that $S(4)$ is true and so on. (What we present here is only an intuitive argument for the validity of PMI. A rigorous justification of PMI requires an appeal to the axioms of **N**.)

Analogously, we might imagine an infinite staircase. Evidently, we can climb the staircase indefinitely if we know that (i) we can step on the first step and (ii) whenever we are on a given step, we can climb to the next step.

Henceforth, we refer to the first step of PMI, the assertion that $S(1)$ is true, as the *basis step* of the proof by mathematical induction, and the second step, the proof that for each positive integer k, $S(k)$ implies $S(k+1)$, as the *inductive step* of the proof by mathematical induction. In most, but not all, proofs by mathematical induction, the basis step is the easier of the two steps to establish. In proving that the inductive step holds, two general strategies are usually used. The first is to produce a direct proof that $S(k)$ implies $S(k+1)$. The second strategy is to reduce $S(k+1)$ back to $S(k)$. Some examples will illustrate these two strategies.

Example 1 Show that for each positive integer n, $1 + \cdots + n = n(n+1)/2$.

Proof. For each positive integer n, let $S(n)$ be the statement $1 + \cdots + n = n(n+1)/2$.

Basis step: $S(1)$ is the statement: $1 = 1(1+1)/2$. Thus, $S(1)$ is true.

Inductive step: We suppose that $S(k)$ is true and prove that $S(k+1)$ is true. Thus, we assume that

$$1 + \cdots + k = k(k+1)/2$$

and prove that $1 + \cdots + k + (k+1) = (k+1)(k+1+1)/2$. If we add $k+1$ to both sides of the equality in $S(k)$, then, on the left side of the sum, we obtain the left side of the equality in $S(k+1)$. Our hope is that the right of the sum equals the right side of $S(k+1)$. Let us check: Adding $k+1$ to both sides of $1 + \cdots + k = k(k+1)/2$, we see that

$$1 + \cdots + k + (k+1) = k(k+1)/2 + (k+1)$$
$$= (k+1)(k/2 + 1)$$
$$= (k+1)(k/2 + 2/2)$$
$$= (k+1)(k+2)/2$$
$$= (k+1)(k+1+1)/2.$$

Hence, if $S(k)$ is true, then $S(k+1)$ is true.

Therefore, by PMI, $1 + \cdots + n = n(n+1)/2$ for each positive integer n. ∎

Example 2 Suppose x is a real number with $x \neq 1$. Show: For any positive integer n, $1 + x + \cdots + x^n = (1 - x^{n+1})/(1 - x)$.

Proof. For each positive integer n, let $S(n)$ be the statement: $1 + x + \cdots + x^n = (1 - x^{n+1})/(1 - x)$ for $x \neq 1$.

Basis step: $S(1)$ reads: $1 + x = (1 - x^2)/(1 - x)$, which is true for $x \neq 1$.

Inductive step: Suppose $S(m)$ is true. We prove $S(m+1)$, i.e., we prove that $1 + x + \cdots + x^{m+1} = (1 - x^{m+1+1})/(1 - x)$. (Note that the use of m instead of k is only a symbolic change. We are proving that if any one statement on the list is true, then the next statement on the list is also true.) Since $S(m)$ is true, $1 + x + \cdots + x^m = (1 - x^{m+1})/(1 - x)$. Thus, adding x^{m+1} to both sides yields:

$$1 + x + \cdots + x^m + x^{m+1} = (1 - x^{m+1})/(1 - x) + x^{m+1}$$
$$= (1 - x^{m+1})/(1 - x) + x^{m+1}(1 - x)/(1 - x)$$
$$= (1 - x^{m+1} + x^{m+1} - x^{m+2})/(1 - x)$$
$$= (1 - x^{m+1+1})/(1 - x).$$

Therefore, $S(m)$ implies $S(m+1)$.

Since both the basis step and the inductive step hold, $1 + x + \cdots + x^n = (1 - x^{n+1})/1 - x)$ for each positive integer n and every real number $x \neq 1$. ∎

The formula given in Example 2 is one of the most important in all of mathematics. The sum on the left side is called the *finite geometric sum with ratio* x. The expression on the right side is called a *closed form* for the geometric sum, meaning that the right side does not involve a sum of n terms. Also note that the formula for $1 + 2 + \cdots + n = n(n+1)/2$ gives a closed form for the sum of the first n positive integers.

Our next example requires us to combine some preliminary investigation and conjecture with an inductive proof.

Example 3   Find the sum of the first n odd integers where n denotes a positive integer.

Let a_n denote the sum of the first n odd integers. Then $a_n = 1 + 3 + \cdots + (2n - 1)$. Our plan is to calculate a_n for several values of n, look for a pattern in the values of a_n, conjecture a formula for a_n on the basis of these observations, and prove the conjecture by mathematical induction.

We form a table to record the values of a_n:

n	1	2	3	4	5	6	7
a_n	1	4	9	16	25	36	49

A clear pattern appears: For each positive integer n, $a_n = 1 + 3 + \cdots + (2n - 1) = n^2$. In a more prosaic way, this result states that the sum of the first n odd integers is n squared.

We can prove this statement by induction. Let $S(n)$ be the statement: For each positive integer n, $1 + \cdots + (2n - 1) = n^2$.

Basis step: $S(1)$ holds since $1 = 1^2$.

Inductive step: Suppose $S(n)$ is true. We must show that $S(n + 1)$ is true. Thus we must show that

$$1 + \cdots + (2(n + 1) - 1) = (n + 1)^2.$$

Since $S(n)$ is true,

$$1 + \cdots + (2n - 1) = n^2.$$

Hence

$$1 + \cdots + (2n - 1) + (2(n + 1) - 1) = n^2 + (2(n + 1) - 1)$$

$$= n^2 + 2n + 2 - 1$$

$$= n^2 + 2n + 1 = (n + 1)^2.$$

Therefore, $S(n + 1)$ is true.

The inductive proof is now complete.   ∎

In the inductive steps of the three examples presented thus far, we have established $S(k + 1)$ by arguing directly from $S(k)$. Now we present an example in which it is not at all obvious how to manipulate $S(k)$ to produce $S(k + 1)$, but where it is natural to reduce $S(k + 1)$ back to $S(k)$.

Example 4   Prove that for each positive integer n, $|\sin(nx)| \le n(\sin(x))$ for $0 \le x \le \pi$. (Here, $|u|$ is the absolute value of the real number u: $|u| = u$ if $u \ge 0$ and $|u| = -u$ if $u < 0$.)

Proof. For each positive integer n, define $S(n)$: $|\sin(nx)| \leq n\sin(x)$ for $0 \leq x \leq \pi$.

Basis step: We must show that $|\sin(1 \cdot x)| = |\sin(x)| \leq 1 \cdot \sin(x) = \sin(x)$ for $0 \leq x \leq \pi$. But for $0 \leq x \leq \pi$, $\sin(x) \geq 0$ hence $|\sin(x)| = \sin(x)$.

Inductive step: We show that for each positive integer k, $S(k)$ implies $S(k+1)$. Therefore, we assume that for $0 \leq x \leq \pi$, $|\sin(kx)| \leq k\sin(x)$ and deduce that $|\sin(k+1)x| \leq (k+1)\sin(x)$.

While it is easy to obtain $(k+1)\sin(x)$ from $k \cdot \sin(x)$ and vice versa, it is not clear how to build $|\sin(k+1)x|$ from $|\sin(kx)|$. But the addition formula for the sine function, $\sin(u+v) = (\sin u)(\cos v) + (\cos u)(\sin v)$, can help us express $\sin(k+1)x$ in terms of $\sin(kx)$. Thus we will attempt to manipulate the left side of $S(k+1)$ to the point where we can use statement $S(k)$. From the addition formula for the sine function,

$$|\sin(k+1)x| = |\sin(kx + x)|$$
$$= |\sin(kx) \cdot \cos(x) + \cos(kx) \cdot \sin(x)|$$
$$\leq |\sin(kx) \cdot \cos(x)| + |\cos(kx) \cdot \sin(x)|$$

since for all real numbers, u and v, $|u+v| \leq |u| + |v|$. (This is the so-called *triangle inequality*.) Moreover, since $|u \cdot v| = |u| \cdot |v|$ for all real u and v and since $|\cos(u)| \leq 1$ for all real u, we have

$$|\sin(k+1)x| \leq |\sin(kx) \cdot \cos(x)| + |\cos(kx) \cdot \sin(x)|$$
$$\leq |\sin(kx)| \cdot |\cos(x)| + |\cos(kx)| \cdot |\sin(x)|$$
$$\leq |\sin(kx)| + |\sin(x)|.$$

Recall that we are assuming that $S(k)$ holds. Thus, $|\sin(kx)| \leq k \cdot \sin(x)$. Also, since $0 \leq x \leq \pi$, $|\sin(x)| = \sin(x)$. It follows that

$$|\sin(k+1)x| \leq |\sin(kx)| + |\sin x|$$
$$\leq k \cdot \sin(x) + \sin(x) \leq (k+1) \cdot \sin(x).$$

Hence, the inductive step is established.

By PMI, $|\sin(nx)| \leq n \cdot \sin(x)$ for each positive integer n and each real number x such that $0 \leq x \leq \pi$. ∎

Let us recap the basic strategy in the last argument. We took an important portion of the inequality that constituted statement $S(k+1)$, namely $|\sin(k+1)x|$, and manipulated it (using properties of the sine, cosine, and absolute value functions) to a point where statement $S(k)$ could be used. In this sense we "reduced" $S(k+1)$ to $S(k)$. This kind of reduction argument stands in contrast to the proofs of inductive steps in Examples 1–3. In those examples, statement $S(k+1)$ was "constructed" from statement

$S(k)$: Statement $S(k)$ is manipulated in an appropriate way to obtain $S(k+1)$. Both of these techniques are used frequently in proofs of the inductive step.

We emphasize that in the inductive step, $S(k + 1)$ is always deduced from statement $S(k)$. Such was the case in each of Examples 1–3. In some cases it is appropriate to begin with statement $S(k)$ and to conclude that statement $S(k + 1)$ holds after a sequence of intermediate steps. In other cases it is more appropriate to use statement $S(k)$ at some intermediate point in the argument that establishes statement $S(k+1)$. Whether or not to begin with statement $S(k)$ is a matter of tactics. But the *logic* of the inductive step is the same in every case.

Our final illustration of PMI yields one of the most important properties of the system of integers, the Division Algorithm. The proof that follows is rather sophisticated since the statement of the Division Algorithm involves two integer variables instead of only one integer variable as all the previous examples have.

Theorem 1   **Division Algorithm.** *Let a and b be positive integers. There exist integers q and r such that $a = b \cdot q + r$, where $0 \leq r \leq b - 1$.*

Remark: The statement asserts that b can be "taken out" of a several times (q times to be exact) in such a way that the remainder is less than b but is at least 0. Here is a quick example: Let $a = 61$ and $b = 13$. Then $61 = 13 \cdot 4 + 9$. Here, $q = 4$ and $r = 9$.

Proof. For each positive integer a, let $S(a)$ be the statement: For each positive integer b there exist integers q and r such that $a = b \cdot q + r$ where $0 \leq r \leq b - 1$.

Basis step: To prove $S(1)$ we let b be an arbitrary positive integer, we must find integers q and r such that $1 = b \cdot q + r$ where $0 \leq r \leq b - 1$. We observe that if $b = 1$, then we can satisfy the desired condition by taking $q = 1$ and $r = 0$. If $b > 1$, then we satisfy the conditions by setting $q = 0$ and $r = 1$.

Inductive step: Suppose $S(a)$ is true. We show $S(a+1)$ is true. Because $S(a)$ is true, there exist integers q and r such that $a = b \cdot q + r$ where $0 \leq r \leq b - 1$. To show that $S(a + 1)$ holds, we must find integers q_1 and r_1 such that $a + 1 = b \cdot q_1 + r_1$ where $0 \leq r_1 \leq b - 1$.

We know that $a = b \cdot q + r$ where $0 \leq r \leq b - 1$. Thus $a + 1 = b \cdot q + r + 1$. If $0 \leq r \leq b - 2$, then $1 \leq r + 1 \leq b - 1$, and $a + 1 = b \cdot q_1 + r_1$ where $q_1 = q$, $r_1 = r + 1$, and $0 \leq 1 \leq r_1 \leq b - 1$. On the other hand if $r = b - 1$, then $r + 1 = b$ and

$$a + 1 = bq + r + 1 = bq + b = b \cdot (q + 1) = b \cdot q_1 + r_1$$

where $q_1 = q + 1$ and $r_1 = 0$. Therefore in every possible case, $S(a + 1)$ holds.

The inductive step is thus established. ∎

One minor technical point arises on occasion. In some cases it is natural to label the statements on the list so that the first statement is $S(2)$, $S(5)$, or in general $S(k_0)$ for some integer $k_0 \geq 0$. For instance, the inequality $2^n \geq n^2$ holds for all n such that $n \geq 4$. In such cases we have a slightly more general version of PMI.

Principle of Mathematical Induction (Generalized)

Let k_0 be a nonnegative integer and let $S(k_0), S(k_0 + 1), \ldots, S(n), \ldots$ be a list of statements, one for each integer $n \geq k_0$. Then every statement on the list is true if the following two conditions hold:

   (i) $S(k_0)$ is true.
   (ii) For each integer $k \geq k_0$, if $S(k)$ is true, then $S(k + 1)$ is true.

In addition to PMI, there is a different form of mathematical induction that possesses two significant virtues: It has the same effect as PMI, and it is applicable in many situations in which PMI cannot be so readily used. We refer to this form of mathematical induction as the Principal of Strong Induction (PSI), since the hypothesis in the inductive step of PSI is stronger than that of PMI.

Principle of Strong Induction

Let k_0 be a nonnegative integer and let $S(k_0), S(k_0+1), \ldots S(n), \ldots$ be a list of statements, one for each positive integer greater than or equal to k_0. Then every statement on the list is true if the following two conditions hold:

   (i) $S(k_0)$ is a true statement.
   (ii) For each positive integer $k \geq k_0$, if $S(k_0), S(k_0 + 1), \ldots, S(k)$ are all true, then $S(k + 1)$ is true.

First note that if PSI (i) and (ii) are established, then $S(k)$ is true for each positive integer $k \geq k_0$. For, if (i) is proved, then $S(k_0)$ is true. By (ii) applied in the case $k = k_0, S(k_0 + 1)$ is true. Since $S(k_0)$ and $S(k_0 + 1)$ are true, it follows from PSI (ii) that $S(k_0 + 2)$ is true. Since $S(k_0)$, $S(k_0 + 1)$, and $S(k_0 + 2)$ are true, $S(k_0 + 3)$ is true by PSI (ii). Again by PSI (ii), we deduce that $S(k_0 + 4)$ holds, and so on. Thus, PSI enables us to prove that $S(k)$ is true for each $k \geq k_0$.

We emphasize that PMI and PSI are similar in that both are proof techniques used to show that every statement in an infinite list of statements is true. However, PMI and PSI differ in a significant way. In the inductive step of PSI, one assumes that all of the first k statements on the list are true and deduces that the $(k + 1)$st statement is true. In the inductive step of PMI, $S(k + 1)$ is deduced from $S(k)$ only. Because more is assumed in the inductive step of PSI than in the inductive step of PMI, the hypothesis of PSI (ii) is stronger than the hypothesis of PMI (ii). We can summarize the two forms of induction in the following chart:

	Basis Step	**Inductive Step**
PMI	Show: $S(k_0)$ is true	Show: $S(k) \Rightarrow S(k+1)$

$$\text{PSI} \quad \text{Show: } S(k_0) \text{ is true} \qquad \text{Show: } \left. \begin{array}{l} S(k_0) \\ S(k_0 + 1) \\ \vdots \\ S(k) \end{array} \right\} \Rightarrow S(k+1)$$

The important question remains: When should one use PSI instead of PMI? The basic idea in the inductive step of PMI is that $S(k+1)$ can be deduced from or reduced back to $S(k)$. This is precisely what occurred in Examples 1–4. In some cases, though, $S(k+1)$ cannot be readily deduced from $S(k)$, but can be derived from all of the previous statements $S(k_0), \ldots, S(k)$. In these cases we appeal to PSI in order to conclude that all the statements $S(k_0), \ldots, S(k), \ldots$ are true. Our next example illustrates this point.

Theorem 2   *Each integer greater than 1 is either prime or is a product of primes.*

Proof. For an integer $n \geq 2$, define the statement $S(n)$: n is either prime or is a product of primes.

Basis step: $S(2)$ is true since 2 is a prime number.

Inductive step: Suppose statements $S(2), \ldots, S(n)$ are all true and show that $S(n+1)$ is true. Now without doubt the integer $n+1$ is either prime or is not prime. If $n+1$ is a prime, then $S(n+1)$ is true. If $n+1$ is not prime, then $n+1 = a \cdot b$ where $a, b < n+1$, hence $a, b > 1$. But by hypothesis, $S(a)$ and $S(b)$ are both true. Thus, each of a and b is either prime or is a product of primes. Write $a = p_1 \cdots p_r$ and $b = q_1 \cdots q_s$ where $p_1, \ldots, p_r$ and $q_1, \ldots, q_s$ are all prime. (For example, if $r = 1$, then a is itself prime.) Therefore,

$$n + 1 = a \cdot b = (p_1 \cdots p_r) \cdot (q_1 \cdots q_s)$$

and $a \cdot b$ is a product of primes.

By PSI, the proof is complete.   ∎

In the inductive step in the proof of Theorem 2, we used the fact that a and b are expressible as a product of primes to conclude that $n + 1 = a \cdot b$ is a product of primes. It is not at all obvious that this conclusion can be derived from the assumption that n is expressible as a product of primes.

To summarize our work, we have introduced two forms of mathematical induction, PMI and PSI. We described each principle, illustrated each with examples, and discussed when it is more appropriate to use PSI instead of

PMI. We have not, however, addressed the question of when a proof by mathematical induction should be used.

The obvious answer appears to be: Use mathematical induction whenever an infinite list of statements is given such as in Examples 1 or 2. However, problems susceptible to inductive proof do not always come so neatly and clearly packaged. (Theorems 1 and 2 bear witness to this observation.) Thus, when should a proof by mathematical induction be used?

As with most matters involving the use of a proof technique, no definitive answer to this question exists. Nonetheless, *it is appropriate to look for a proof by mathematical induction of any statement that involves an integer variable*. For example, in Theorem 2 the variable is the integer to be factored. The integer variable appearing in the problem statement can then be used to number the statements on the infinite list of statements. Of course, we are not claiming that every statement in which an integer variable appears must be proved by mathematical induction. But as a general policy, induction should be considered as a proof technique for such a statement.

EXERCISES §7

1. Prove: For each positive integer n, $1+4+7+\cdots+(3n-2) = (3n^2-n)/2$.
2. Prove: For each positive integer n, $1+5+9+\cdots+(4n-3) = 2n^2 - n$.
3. Prove: For each positive integer n and each real $x \geq -1$, $(1+x)^n \geq 1 + nx$.
4. Prove: $1^2 + 3^2 + \cdots + (2n-1)^2 = (4n^3 - n)/3$ for each positive integer n.
5. Show: $n^2 \leq 2^n$ for each integer $n \geq 4$.
6. Prove: $(1/5)x^5 + (1/3)x^3 + (7/15)x$ is an integer for each positive integer x.
7. Prove: For each positive integer n,

$$\frac{1}{n+1} + \cdots + \frac{1}{2n} = (1 - 1/2) + (1/3 - 1/4) + \cdots + \left(\frac{1}{2n-1} - \frac{1}{2n}\right).$$

8. (Generalized distributive law). Let $a, b_1, \ldots, b_n$ be real numbers. Show: $a \cdot (b_1 + \cdots + b_n) = a \cdot b_1 + \cdots + a \cdot b_n$.
9. Prove: For any integer $n \geq 2$, the product of n odd integers is an odd integer.
10. For $n \geq 1$ let $s_n = 1 + (8 \cdot 1 + \cdots + 8 \cdot n)$.
    (a) Compute s_1, s_2, s_3, s_4, and s_5.
    (b) Conjecture a closed formula for s_n for an arbitrary value of n.
    (c) Prove your conjecture by mathematical induction.

11. Let $a_n = 1/(1 \cdot 2) + 1/(2 \cdot 3) + \cdots + 1/(n(n+1))$ for each positive integer n.
    (a) Compute a_1, a_2, a_3, a_4, and a_5.
    (b) Conjecture a closed formula for a_n for an arbitrary value of n.
    (c) Prove your conjecture by mathematical induction.
    (d) Show that the infinite series $\sum\limits_{k=1}^{\infty} 1/k(k+1)$ converges and determine its sum.

12. Let $b_n = 1/3 + 1/15 + \ldots + 1/(4n^2 - 1)$.
    (a) Compute b_1, b_2, b_3, b_4, and b_5.
    (b) Conjecture a closed formula for b_n.
    (c) Prove your conjecture by mathematical induction.

13. Let $c_n = 1^3 + \cdots + n^3$.
    (a) Compute c_1, c_2, c_3, c_4, and c_5.
    (b) Conjecture a closed formula for c_n.
    (c) Prove your conjecture.

14. For each positive integer k, define the integer $k!$ (called "k factorial") by $k! = k \cdot (k-1) \cdot \cdots \cdot 2 \cdot 1$. Let $e_n = 1 \cdot 1! + 2 \cdot 2! + 3 \cdot 3! + \cdots + n \cdot n!$. Conjecture and prove a closed formula for e_n.

15. Let $d_n = \sum_{k=1}^{n} k/(k+1)!$.
    (a) Compute d_1, d_2, d_3, d_4, and d_5.
    (b) Conjecture a closed formula for d_n.
    (c) Prove your conjecture.

16. Conjecture and prove a closed formula for the product $(1 - 1/4) \cdot (1 - 1/9) \cdot \cdots \cdot (1 - 1/n^2)$.

17. Let f be a function that has derivatives of every order at each real number x; i.e., $f'(x), f''(x), \ldots, f^{(n)}(x) \ldots$ exist for each positive integer n and each real number x. Let $g(x) = x \cdot f(x)$. Conjecture and prove a formula for $g^{(n)}(x)$ that expresses $g^{(n)}(x)$ in terms of $f^{(n)}(x)$ and x.

18. (a) Prove: For every positive integer n, 3 divides $n^3 - n$.
    (b) Prove: For every positive integer n, 5 divides $n^5 - n$.
    (c) Formulate a statement that contains the statements in (a) and (b) as special cases.

19. Define a sequence of positive integers m_n, $n \geq 1$, by $m_1 = 2$, and for $n \geq 2$, $m_n = m_{n-1} \cdot (m_{n-1} - 1) + 1$.
    (a) Compute m_2, m_3, and m_4.
    (b) Show that for $n \geq 1$, $m_{n+1} = (m_1 \cdot \cdots \cdot m_n) + 1$.

20. Male bees, which have a mother but no father, hatch from unfertilized eggs. Female bees hatch from fertilized eggs. How many ancestors does a male bee have in the tenth generation back? How many of these ancestors are male?

21. Prove that every positive integer has a base 2, or *binary*, expansion. Specifically prove that if n is a positive integer, then there exist integers $a, a_1, \ldots, a_k$ such that each $a_i = 0$ or 1 and $n = a_0 + a_1 \cdot 2 + \cdots + a_k \cdot 2^k$.

Section 8
CASE ANALYSIS

The final method of proof to be discussed is proof by *case analysis*. In a case analysis proof of a given statement, the entire argument is divided into a collection of cases. Each of the cases is then resolved, thereby establishing the original statement.

For example, in proving some statement involving a single integer variable, call it x, one might consider two cases: (i) $x \geq 0$ and (ii) $x < 0$. Or in proving a statement involving a real number x, one can analyze the following cases: (i) $x =$ integer, (ii) $x =$ rational, (iii) $x =$ arbitrary real.

We distinguish two types of case analysis proofs: *divide-and-conquer* and *bootstrap*. In a divide-and-conquer proof, the original problem is divided into a number of separate cases that are established (or conquered) independently of each other. In a bootstrap proof, the original problem is divided into a sequence of cases in which a given case after the first one is established with the use of some or all of the previous cases. We now illustrate each of these types of case analysis proofs. We begin with an example of a divide-and-conquer argument.

Example 1 Find all real solutions to the inequality $|x - 1| < |x - 3|$.

Recall that
$$|x| = \begin{cases} x & \text{if } x \geq 0 \\ -x & \text{if } x < 0. \end{cases}$$

Thus,

$$|x - 1| = \begin{cases} x - 1 & \text{if } x - 1 \geq 0, \text{ i.e., if } x \geq 1 \\ -(x - 1) = 1 - x & \text{if } x - 1 < 0, \text{ i.e., if } x < 1 \end{cases}$$

and

$$|x - 3| = \begin{cases} x - 3 & \text{if } x \geq 3 \\ 3 - x & \text{if } x < 3. \end{cases}$$

Thus we consider three cases: 1. $x \geq 3$, 2. $1 \leq x < 3$, and 3. $x < 1$.

Case 1. $x \geq 3$. Then $|x-1| = x-1$ and $|x-3| = x-3$ and the original inequality reduces to $x - 1 < x - 3$, which holds if and only if $-1 < -3$. Thus $|x - 1| < |x - 3|$ holds for no $x \geq 3$.

Case 2. $1 \leq x < 3$. In this case $|x - 1| = x - 1$ and $|x - 3| = 3 - x$ and the inequality $|x - 1| < |x - 3|$ becomes $x - 1 < 3 - x$, which holds if

71

and only if $2x < 4$. Therefore, for $1 \leq x < 3$, $|x - 1| < |x - 3|$ is valid if $1 \leq x < 2$.

Case 3. $x < 1$. Then the given inequality becomes $1 - x < 3 - x$ which is equivalent to $1 < 3$. Thus $|x - 1| < |x - 3|$ holds for all $x < 1$.

From Cases 1–3 we conclude that $|x - 1| < |x - 3|$ holds if and only if $x < 2$.

Notice that the original problem is divided into three cases, or subproblems. These cases are independent in that the solution to any particular case does not depend on the solutions to the other cases. Thus, the order in which the cases are considered is irrelevant. Also, it is perhaps obvious but important to note that every possible value of x is handled by one of the cases.

Example 2   Let us prove the triangle inequality:

$$\text{For all } x, y \text{ in } \mathbf{R}, \ |x + y| \leq |x| + |y|.$$

Again we consider several separate cases.

Case 1. $x \geq 0$ and $y \geq 0$. Then $x + y \geq 0$ and $|x+y| = x+y = |x|+|y|$.

Case 2. $x \geq 0$ and $y < 0$. Then $|x| = x$ and $|y| = -y$. The value of $|x + y|$ depends on whether $x + y \geq 0$ or $x + y < 0$. Let us therefore consider subcases:

Subcase 1. $x + y \geq 0$. Then

$$|x + y| = x + y \leq x + (-y) \qquad \text{since } y < 0 < -y$$
$$= |x| + |y|.$$

Subcase 2. $x + y < 0$. Then, since $-x \leq x$ for $x \geq 0$,

$$|x + y| = -(x + y) = (-x) + (-y) \leq x + (-y) = |x| + |y|.$$

Case 3. $x < 0$ and $y \geq 0$. This case is handled by an argument similar to that given in Case 2.

Case 4. $x < 0$ and $y < 0$. Then $x + y < 0$ and $|x + y| = -(x + y) = (-x) + (-y) = |x| + |y|$.

Next we illustrate a bootstrap-type case analysis proof.

Example 3   Suppose f is a function on the real numbers with the property that $f(x + y) = f(x) + f(y)$ for all real numbers x and y. (We call this the additive property of f.) Prove that for all real numbers x and all rational numbers r

$$f(r \cdot x) = r \cdot f(x).$$

Proof. The bootstrap proof is divided into five cases:
1. $r = 0$,

2. $r =$ positive integer,
3. $r =$ negative integer,
4. $r = 1/n$ where n is a nonzero integer,
5. $r = m/n$ where m is an integer and n is a nonzero integer.

Case 1. We show $f(0) = 0$. $f(0) = f(0 + 0) = f(0) + f(0)$, hence $f(0) = 0$. Thus for any real x, $f(0 \cdot x) = f(0) = 0 \cdot f(x)$.

Case 2. We show $f(n \cdot x) = n \cdot f(x)$ for any real x and any positive integer n. To demonstrate this claim we use PMI. For any positive integer n, let $S(n)$ be the statement: $f(n \cdot x) = n \cdot f(x)$ for any real x.

Basis step. $f(1 \cdot x) = f(x) = 1 \cdot f(x)$ for any real x. Thus $S(1)$ holds.
Inductive step. We show $S(n)$ implies $S(n + 1)$. Suppose $f(n \cdot x) = n \cdot f(x)$ for all real x. We show $f((n + 1) \cdot x) = (n + 1) \cdot f(x)$ for all real x. By the additive property of f,

$$f((n + 1) \cdot x) = f(n \cdot x + x) = f(nx) + f(x)$$
$$= n \cdot f(x) + f(x) \qquad \text{by inductive hypothesis}$$
$$= (n + 1) \cdot f(x).$$

Therefore, by PMI $f(n \cdot x) = n \cdot f(x)$ for all real x and all positive integers n.

Case 3. Let n be a negative integer. We show $f(n \cdot x) = n \cdot f(x)$. By Case 1 and the additive property of f,

$$0 = f(0) = f(n \cdot x + (-n \cdot x)) = f(nx) + f(-n \cdot x).$$

Since $-n$ is a positive integer, $f(-n \cdot x) = -n \cdot f(x)$ by Case 2. Hence

$$f(n \cdot x) + f(-n \cdot x) = f(n \cdot x) - n \cdot f(x) = 0$$

and

$$f(n \cdot x) = n \cdot f(x).$$

We now know that $f(n \cdot x) = n \cdot f(x)$ for all real x and any integer n.

Case 4. We show that if $r = 1/n$ where n is a nonzero integer, then $f(r \cdot x) = r \cdot f(x)$ for any real x. We have

$$f(x) = f(n \cdot (1/n) \cdot x) = n \cdot f((1/n) \cdot x)$$

by Case 2 or Case 3.
Thus $f((1/n) \cdot x) = 1/n \cdot f(x)$.

Case 5. Let r be any rational number and x any real number. Then $r = m/n$ where m and n are integers and $n \neq 0$. Thus by Cases 2, 3, and 4,

$$f(r \cdot x) = f(m/n \cdot x) = f(m \cdot x/n) = m \cdot f(x/n)$$

$$= m/n \cdot f(x)$$

$$= r \cdot f(x). \quad \blacksquare$$

Observe that every case in this proof after the first two refers back to earlier cases. In general, a bootstrap proof is built up in a sequence of layers with a given layer relying on previous layers. Here is one more example.

Example 4 Show that an angle that is inscribed in a circle has measure equal to half the central angle that subtends the same arc.

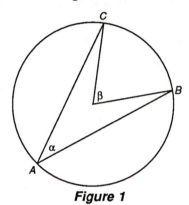

Figure 1

With reference to Figure 1, we want to prove that $\alpha = \frac{1}{2}\beta$.

Case 1. Suppose that one of the line segments AB or AC contains the center O of the circle. See Figure 2.

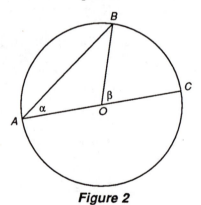

Figure 2

Then $180° - \beta = \angle AOB = 180° - \angle OAB - \angle OBA = 180° - \alpha - \alpha$ since $\angle OAB = \alpha$ and triangle OAB is isosceles. Therefore $\beta = 2\alpha$.

Case 2. Suppose that the center O is interior to $\angle BAC$. See Figure 3.

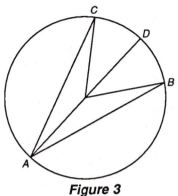

Figure 3

Extend the line segment AO to a segment AD where D lies on the circle. Then by Case 1,

$$\alpha = \angle CAB = \angle CAD + \angle BAD$$

$$= \frac{1}{2}\angle COD + \frac{1}{2}\angle BOD$$

$$= \frac{1}{2}\angle COB = \frac{1}{2}\beta.$$

Case 3. The center O is exterior to $\angle BAC$. See Figure 4.

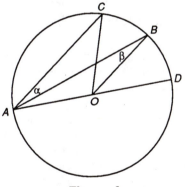

Figure 4

Again extend the segment AO to AD where D lies on the circle. Then

$$\alpha = \angle CAB = \angle CAD - \angle BAD$$

$$= \frac{1}{2}\angle COD - \frac{1}{2}\angle BOD$$

$$= \frac{1}{2}\angle COB = \frac{1}{2}\beta.$$

EXERCISES §8

1. Solve the inequality $|x + 1| < |x - 1|$.
2. Solve the inequality $|x + 1| < |x^2 - 1|$.
3. (a) Show that if a is an integer, then there exists an integer q such that $a = 3 \cdot q + r$ where $r = 0, 1,$ or 2. (You will not need a case analysis; just use the Division Theorem.) Now use this result together with a proof by contradiction and case analysis in part (b).
   (b) Show that if 3 divides a product $a \cdot b$ where a and b are integers, then either 3 divides a or 3 divides b.
4. Investigate the following extension of Example 4. Let the two chords of a circle meet at a point P on or inside the circle. Let $\alpha = \angle BPD$ and let β and γ be the central angles that subtend the arcs BD and AC respectively. Express α in terms of β and γ. (Note that Example 4 is the special case of this situation in which $A = C = P$.)

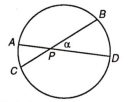

5. Can one form a ten digit integer by putting a digit between 0 and 9 in the empty boxes in the given table as follows: the digit in the box labeled 0 indicates the number of times 0 appears in the number, the digit in box 1 indicates the number of times 1 appears in the number, etc.?

0	1	2	3	4	5	6	7	8	9

(For instance, if 9 is placed in box 0, then the remaining boxes must be filled with 0 in order that the nine zeros actually appear. In this case, however, we reach a contradiction since box 9 cannot contain 0 as at least one 9 appears in the number. Thus the desired ten-digit number cannot have 9 as its leftmost digit.) (Hint: Consider the possible ways of filling the box labeled 0.) How many such numbers exist?

6. The natural logarithm function, log, has the property that $\log(xy) = \log(x) + \log(y)$ for all positive reals x and y. Show that $\log(x^r) = r \log(x)$ for all positive real numbers x and all rational numbers r.

7. Let f be a function on the real numbers such that $f(0) \neq 0$ and $f(x + y) = f(x) \cdot f(y)$ for all real x and y. Conjecture and prove a formula for $f(r \cdot x)$ valid for all rational numbers r and real numbers x.

Section 9
QUANTIFICATION
AND
COUNTEREXAMPLES

At this time it is perhaps appropriate to make some general comments on the logical structure of mathematical proofs of quantified statements. The remarks we are about to record have been implicit throughout this unit, but it is useful to state them explicitly at this point.

To prove a statement of the form "for all x, $P(x)$," one must show that for each x in the given domain of the problem (e.g., for each real number x or for each integer x), the statement $P(x)$ is valid. This task can be carried out in several ways: either via a direct proof, an indirect proof (contraposition or contradiction), mathematical induction, or a case analysis. Several statements of the type "for all x, $P(x)$" were proved in this section.

To prove a statement of the form "there exists x such that $P(x)$," one can choose one of two methods. The first is to construct directly an object x_0 such that the statement $P(x_0)$ is valid. For instance, to show that there exists a pair (x, y) of real numbers satisfying the system of equations

$$2x + y = 2$$
$$4x - 2y = 0,$$

one can simply show that the pair $(1/2, 1)$, (derived by simultaneously solving both equations), does indeed satisfy both equations.

A second method of demonstrating "there exist x such that $P(x)$" is an argument by contradiction: Assume that the negation of this statement holds and derive a contradiction. Thus, we suppose that it is not the case that "there exists x such that $P(x)$." In other words, we suppose that "for all x, not-$P(x)$." Then from this statement, one derives a contradiction that implies that the original statement is valid, and therefore, an object x satisfying $P(x)$ does indeed exist. For example, the proof that there exists a real number u such that $u^2 = 2$ is, if slightly recast, an illustration of this method. (See Section 6, Theorem 3.) If one assumes that there is no real number whose square is 2, then it follows that the least upper bound u of the set $A = \{x \in \mathbf{R} \mid x^2 < 2\}$ has the property that $u^2 \not< 2$, $u^2 \neq 2$, $u^2 \not> 2$.

This violation of the Trichotomy Law means that there exists a square root of 2 in $\mathbf{R}$.

Example 1 Let a be a positive real number. Prove: There exists a real number x_0 such that $x_0^2 = a$.

Proof. We use properties of continuous functions, particularly the Intermediate Value Theorem, to derive a contradiction. Suppose that for all real x, $x^2 \neq a$. Consider the function $f(x) = x^2 - a$. Since $x^2 \neq a$ for all real numbers x, $f(x) = x^2 - a \neq 0$ for all real numbers x. Also f is continuous on $\mathbf{R}$. Note that $f(a+1) = (a+1)^2 - a = a^2 + a + 1 > 0$ while $f(0) = -a < 0$. Thus f is a continuous function taking on both positive and negative values, yet $f(x) \neq 0$ for all x. Thus we have a violation of the Intermediate Value Theorem. This contradiction completes the argument. ∎

Finally, let us discuss how the statement "for all x, $P(x)$" can be *disproved*. As noted above, to prove such a statement, one must show that for each x, the statement $P(x)$ is valid. Thus, to disprove "for every x, $P(x)$," one must show that "there exists x_0 such that not-$P(x_0)$" is valid, i.e., one must find at least one x_0 such that $P(x_0)$ does not hold. An object x_0 with this property is called a *counterexample* to $P(x)$. Thus disproving "for every x, $P(x)$" amounts to finding a counterexample to $P(x)$.

Example 2 Consider the statement: For every 2×2 matrix A with real number entries, if $A \neq \begin{pmatrix} 0 & 0 \\ 0 & 0 \end{pmatrix}$, then there exists a 2×2 matrix B such that $AB = I = \begin{pmatrix} 1 & 0 \\ 0 & 1 \end{pmatrix}$.

The given statement is true if one assumes that A and B are real numbers (i.e., A and B are 1×1 matrices). But for 2×2 matrices, matters are more complicated. In fact the matrix $A = \begin{pmatrix} 1 & 0 \\ 0 & 0 \end{pmatrix}$ is not the zero matrix, yet for any 2×2 matrix $B = \begin{pmatrix} a & b \\ c & d \end{pmatrix}$, $AB = \begin{pmatrix} a & b \\ 0 & 0 \end{pmatrix}$. Thus, for each 2×2 matrix B, $AB \neq \begin{pmatrix} 1 & 0 \\ 0 & 1 \end{pmatrix} = I$, and $A = \begin{pmatrix} 1 & 0 \\ 0 & 0 \end{pmatrix}$ provides a counterexample to the given statement.

As this example illustrates, to find a counterexample to an implication involving a variable, "if $P(x)$ then $Q(x)$," one must find a specific value of x, call it x_0, for which $P(x_0)$ is true and $Q(x_0)$ is false. In Example 2, the matrix $A = \begin{pmatrix} 1 & 0 \\ 0 & 0 \end{pmatrix}$ is a 2×2 matrix with real entries such that $A \neq \begin{pmatrix} 0 & 0 \\ 0 & 0 \end{pmatrix}$ (hence $P(A)$ is true) and for each 2×2 matrix B, $AB \neq I$ (and $Q(A)$ is false).

Once a counterexample to a statement has been obtained, our work is not finished. For we should seek ways of modifying the original statement so as to obtain a true statement. Specifically, we can try to determine some additional conditions that, when added to the hypotheses of the statement, allow us to deduce the conclusion. For instance, in Example 2, if we modify the hypothesis by insisting that the determinant of A is not 0 rather than $A \neq \begin{pmatrix} 0 & 0 \\ 0 & 0 \end{pmatrix}$, then the resulting statement is true:

> If A is a 2×2 matrix with real entries such that the determinant of A is not 0, then there is a matrix B such that $AB = \begin{pmatrix} 1 & 0 \\ 0 & 1 \end{pmatrix}$.

Let us consider an example that illustrates a type of exercise that will appear periodically throughout this text.

Example 3   Prove or disprove and salvage, if possible: For any integer $n > 1$, $\sqrt{n}$ is irrational.

There are two available options: (1) Prove the given statement or (2) disprove the statement by giving a counterexample, then modify the statement and prove the new statement. Since the given statement involves an integer variable, an approach through mathematical induction is suggested. The first case to consider is $n = 2$ and by Example 4 of Section 5, $\sqrt{2}$ is indeed irrational. This result constitutes the basis step in a proof by mathematical induction of the statement $S(n) : \sqrt{n}$ is irrational. Rather than trying to prove the inductive step, we can search for more evidence in support of the statement. For instance, $\sqrt{3}$, $\sqrt{6}$, and $\sqrt{8}$ can all be proved to be irrational. But by this point a counterexample has perhaps surfaced: $\sqrt{4} = 2$, which is a rational number; moreover, for any positive integer n, $\sqrt{n^2} = n$, which is rational. Thus, we must modify the original statement.

There are at least two ways of approaching this task. The first is to consider other specific cases ($\sqrt{5}$, $\sqrt{7}$, $\sqrt{11}$, or $\sqrt{12}$, for example), then formulate a conjecture based on the finding in these cases. The second approach is simply to revise the original statement so as to exclude the counterexamples observed above ($\sqrt{n^2} = n$). In the latter case we come up with the statement:

(∗)   If n is a positive integer that is not the square of another integer, then $\sqrt{n}$ is irrational.

Even though (∗) contains an integer variable, that variable ranges only over the nonsquare positive integers. Thus, a proof by mathematical induction does not seem feasible. What kind of proof should be attempted? Is the result true? Does (∗) require further modification?

EXERCISES §9

Give a counterexample to each of the following statements.

1. The sum of two irrational numbers is irrational.
2. The product of two irrational numbers is irrational.

3. If a and b are positive integers and $a \cdot b$ is a perfect square, then a and b are perfect squares.
4. The square root function has the additive property.

In each of the following exercises, write the given statement as an implication (a) without using variables and (b) using variables and using quantifiers.

5. The product of an even integer and an odd integer is an even integer.
6. The square root of an irrational number is irrational.
7. The cube root of 2 is irrational.
8. The derivative of a constant function is 0.
9. The determinant of an invertible matrix is nonzero.
10. The product of two irrational numbers is irrational.
11. The rank of a singular $n \times n$ matrix is less than n.
12. The sum of two irrational numbers is irrational.
13. For a function to be constant, it is sufficient that its derivative be identically zero.

Section 10

SETS
AND
SET OPERATIONS

We conclude this unit by extending our study of sets and set operations. We begin with the basic properties of the set operations of union, intersection, difference, and complement. We then consider the power set of a set and conclude with the Cartesian product of sets. Sets in themselves happen to be interesting objects and are therefore worthy of study. Moreover, sets and the theory of sets have assumed a central position in mathematics in this century. For example, many mathematicians maintain that any mathematical entity should be describable as a set, and so, they would say, anyone wishing to study modern mathematics should become conversant with the vocabulary and concepts of set theory. In addition to these reasons, we will use this discussion of sets as a vehicle to illustrate the various proof techniques presented in this unit.

PROPERTIES OF OPERATIONS

What are some of the properties of the set operations of union, intersection, difference, and complement? As a guide in this inquiry, let us recall from Section 6 some of the properties of the operations of addition and multiplication of real numbers. As usual, let $+$ and $\cdot$, respectively, denote these operations.

Perhaps the most familiar properties are the commutative and associative laws: for all $a, b, c \in \mathbf{R}$,

(i) Commutative laws
   (a) $a + b = b + a$,
   (b) $a \cdot b = b \cdot a$
(ii) Associative laws
   (a) $(a + b) + c = a + (b + c)$
   (b) $(a \cdot b) \cdot c = a \cdot (b \cdot c)$

The two operations are connected by the distributive law of multiplication over addition:

(iii) Distributive law
$$a \cdot (b + c) = a \cdot b + a \cdot c$$

81

Also, the number 0 plays a special role with respect to both operations: for each $a \in \mathbf{R}$, $a + 0 = a$ and $a \cdot 0 = 0$.

The operations $\cup$ and $\cap$ possess these properties as well as a few others.

Theorem 1 *Let A, B, and C be any sets.*
- (i)   (a) $A \cup \emptyset = A$.
-       (b) $A \cap \emptyset = \emptyset$.
- (ii) *Idempotent laws*   (a) $A \cup A = A$.
-                           (b) $A \cap A = A$.
- (iii) *Associative laws*   (a) $(A \cup B) \cup C = A \cup (B \cup C)$.
-                            (b) $(A \cap B) \cap C = A \cap (B \cap C)$.
- (iv) *Commutative laws*   (a) $A \cup B = B \cup A$.
-                            (b) $A \cap B = B \cap A$.
- (v)   *Distributive laws*   (a) $A \cup (B \cap C) = (A \cup B) \cap (A \cup C)$.
-                             (b) $A \cap (B \cup C) = (A \cap B) \cup (A \cap C)$.

This theorem suggests that the analogies between the operations of addition and multiplication of numbers and the operations of union and intersection of sets are strong. Commutative and associative laws hold for both the operations. If the empty set is regarded as the set analog of the number 0, then union and addition behave in similar ways. For the union of $\emptyset$ with any other set is that set, while the sum of 0 and any other number is that number. Also multiplication and intersection behave in similar ways. For the product of 0 and any number is 0, while the intersection of $\emptyset$ with any set is $\emptyset$. Thus union seems to be analogous to addition, and intersection seems to be analogous to multiplication.

The analogy, however, is not exact. For neither addition nor multiplication possesses the idempotent property: It is not true that for each real number a, $a + a = a$ and $a \cdot a = a$. And while multiplication of real numbers does distribute over addition, addition does not return the favor: the equality $a + (bc) = (a + b) \cdot (a + c)$ does not hold for all $a, b, c \in \mathbf{R}$.

Where do these remarks leave us? As a heuristic principle, we can maintain and use profitably the idea that union and intersection are analogous to addition and multiplication, respectively. Hence we will continue to use the real number system as a source in our investigation of sets. We cannot, however, thoughtlessly assert that every property of the real number system should translate directly into a property for sets or *vice versa*. Rather we must check each purported analogy carefully before accepting or rejecting it. In carrying out this investigation, we will, of course, be using the standard tools of mathematics that were discussed in Chapter 1: examples to develop experience and to provide counterexamples, general and specific representations to provide various viewpoints and conjectures (in this case of sets, Venn diagrams and digital representations), and proof techniques to provide convincing demonstrations of any conjectures.

Now to the verification of Theorem 1. Proofs of (b) in parts (i) and

(v) will be given here; the proofs of (a) in (ii)–(iv) and (v) will be left as exercises. Although statement (i) is obvious, we present its proof as another example of an element-chasing proof to establish the equality of two sets.

Proof of Theorem 1 (i)(a) To prove that $A \cup \emptyset = A$, we show that $A \cup \emptyset \subseteq A$ and $A \subseteq A \cup \emptyset$.

To prove that $A \cup \emptyset \subseteq A$, we take an arbitrary element x in $A \cup \emptyset$ and show that $x \in A$. If $x \in A \cup \emptyset$, then either $x \in A$ or $x \in \emptyset$. But $\emptyset$ has no elements, hence $x \notin \emptyset$. Thus, $x \in A$. Therefore, $A \cup \emptyset \subseteq A$.

Now we show $A \subseteq A \cup \emptyset$ by taking an arbitrary element $x \in A$ and proving that $x \in A \cup \emptyset$. But if $x \in A$, then either $x \in A$ or $x \in \emptyset$; in other words, if $x \in A$, then $x \in A \cup \emptyset$ and $A \subseteq A \cup \emptyset$.

We conclude that $A \cup \emptyset = A$. ∎

Proof of (i)(b) We show $A \cap \emptyset \subseteq \emptyset$ and $\emptyset \subseteq A \cap \emptyset$.

Because the empty set is a subset of any set (Theorem 1(a)), $\emptyset \subseteq A \cap \emptyset$.

To establish the inclusion $A \cap \emptyset \subseteq \emptyset$, we argue by contradiction. Suppose $A \cap \emptyset \not\subseteq \emptyset$. Hence there exists an element $x \in A \cap \emptyset$ such that $x \notin \emptyset$. However, if $x \in A \cap \emptyset$, then $x \in A$ and $x \in \emptyset$. But since $x \in \emptyset$, $\emptyset$ is not empty, a conclusion that violates the definition of $\emptyset$. Therefore, the assumption that $A \cap \emptyset \not\subseteq \emptyset$ is incorrect. Thus, $A \cap \emptyset \subseteq \emptyset$. Therefore we know that $A \cap \emptyset = \emptyset$. ∎

(Note that the proof by contradiction is appropriate, since by negating the desired conclusion $A \cap \emptyset \subset \emptyset$, we immediately obtain an element $x \in A \cap \emptyset$ that we can work with.)

Proof of (v)(b) We show (1) $A \cap (B \cup C) \subseteq (A \cap B) \cup (A \cap C)$ and (2) $(A \cap B) \cup (A \cap C) \subseteq A \cap (B \cup C)$.

(1) We take an arbitrary $x \in A \cap (B \cup C)$ and show $x \in (A \cap B) \cup (A \cap C)$. Since $x \in A \cap (B \cup C)$, $x \in A$ and $x \in B \cup C$. Thus, if $x \in A \cap (B \cup C)$, then $x \in A$ and either $x \in B$ or $x \in C$. We have then two possible cases:

Case 1. $x \in A$ and $x \in B$.

Case 2. $x \in A$ and $x \in C$.

In Case 1, $x \in A \cap B$; in Case 2, $x \in A \cap C$. Since either Case 1 or Case 2 is true, either $x \in A \cap B$ or $x \in A \cap C$; that is, $x \in (A \cap B) \cup (A \cap C)$. Therefore, $A \cap (B \cup C) \subseteq (A \cap B) \cup (A \cap C)$.

(2) We complete the proof by showing that $(A \cap B) \cup (A \cap C) \subseteq A \cap (B \cup C)$. Let $x \in (A \cap B) \cup (A \cap C)$. We must show that $x \in A \cap (B \cup C)$.

If $x \in (A \cap B) \cup (A \cap C)$, then $x \in A \cap B$ or $x \in A \cap C$. Once again we have a divide-and-conquer type case analysis. Consider two cases:

Case 1. $x \in A \cap B$.

Case 2. $x \in A \cap C$.

In Case 1, $x \in A$ and $x \in B$. Since $x \in B$, either $x \in B$ or $x \in C$. Thus, $x \in A$ and $x \in B \cup C$, which means that $x \in A \cap (B \cup C)$.

In Case 2, $x \in A \cap C$ and an argument identical to that one just presented shows that $x \in A \cap (B \cup C)$. We have proved that $x \in A \cap (B \cup C)$ whenever $x \in (A \cap B) \cup (A \cap C)$, or equivalently that $(A \cap B) \cup (A \cap C) \subseteq A \cap (B \cup C)$.

The proof of Theorem 1 (v)(b) is now complete.   ∎

Questions concerning set difference also arise: What are the basic properties of set difference? What if any are the connections between set difference and the operation of union and intersection and other previously encountered concepts?

Let us investigate one reasonable question: Does set difference distribute over union? Does $A - (B \cup C) = (A - B) \cup (A - C)$ for all sets A, B, and C?

There are several ways to approach this question. One method is to search immediately for a proof of the result. A second is to analyze various examples in an effort either to find a counterexample or to develop an idea for a proof. Finally, one could seek a general "picture" of the problem by scrutinizing Venn diagrams.

Trying to prove a result before being convinced of its truth is at best unwise. So let us consider some special cases. In doing so, it usually pays to take examples that are especially simple. For example, if $A = B = C$, then $A - (B \cup C) = \emptyset = (A - B) \cup (A - C)$. A somewhat more complicated case arises when $A = B$ and $A \neq C$. For example, let us take $A = B = \{1\}$ and $C = \{2\}$. Then $A - (B \cup C) = \{1\} - \{1, 2\} = \emptyset$, while $(A - B) \cup (A - C) = \emptyset \cup \{1\} = \{1\}$. Thus, the conjecture that $A - (B \cup C) = (A - B) \cup (A - C)$ is not true in general.

At this point one can ask at least two questions:

(1) For which sets A, B, C, does $A - (B \cup C) = (A - B) \cup (A - C)$? Ideally, the answer would have the form: $A - (B \cup C) = (A - B) \cup (A - C)$ if and only if _____, where the blank is filled by a statement prescribing properties that A, B, and C satisfy.

(2) Can the statement $A - (B \cup C) = (A - B) \cup (A - C)$ be modified to obtain an assertion that is true for all sets A, B, and C?

Let us investigate the second question and leave the first as an exercise. To do so, we use Venn diagrams as an investigative tool. Remember Venn diagrams have the advantage of being general and easy to manipulate. They have the disadvantage of being imprecise. But at this stage we are only interested in making a conjecture, and hence are willing to subjugate our logical inclinations to our intuition and imagination.

First, a Venn diagram for $A - (B \cup C)$ is drawn (Figure 1).

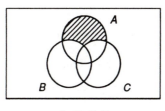

$A - (B \cup C)$

Figure 1

Then, Venn diagrams for $A - B$ and $A - C$ are drawn as in Figure 2.

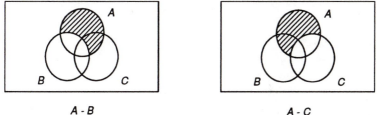

$A - B$ $A - C$

Figure 2

Let us combine these diagrams into one, as shown in Figure 3.

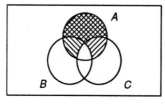

Figure 3

The region with the cross-hatching is exactly $(A - B) \cap (A - C)$. Yet this region appears to coincide with $A - (B \cup C)$. Thus, we conjecture that $A - (B \cup C) = (A - B) \cap (A - C)$. This assertion is indeed true, as the reader can demonstrate with an element-chasing argument.

What about other possible distributive laws, such as those involving $A - (B \cap C)$, $A \cap (B - C)$, and $(A \cup B) - C$? Pause for a moment and investigate each of these sets with Venn diagrams in order to make conjectures concerning them. The answers are recorded in the next theorem.

Theorem 2 *Let A, B, and C be sets.*

    (i) $A \subseteq B$ *if and only if* $A - B = \emptyset$.

    (ii) $A \cap B = \emptyset$ *if and only if* $A - B = A$.

    (iii) $A - (B \cup C) = (A - B) \cap (A - C)$.

    (iv) $A - (B \cap C) = (A - B) \cup (A - C)$.

    (v) $A \cap (B - C) = (A \cap B) - (A \cap C)$.

    (vi) $(A \cup B) - C = (A - C) \cup (B - C)$.

We shall prove parts (i) and (vi). The proofs of parts (ii)–(v) are left as exercises, although special cases of parts (iii) and (iv) will be proved in the next theorem. As an exercise, before trying to prove any of the parts of the theorem, draw Venn diagrams illustrating each statement.

Proof of (i)   We show: (1) If $A \subseteq B$, then $A - B = \emptyset$, and (2) if $A - B = \emptyset$, then $A \subseteq B$.

Case 1. To show that $A \subseteq B$ implies $A - B = \emptyset$, we argue by contraposition. Thus we assume $A - B \neq \emptyset$ and show that $A \not\subseteq B$. Since $A - B \neq \emptyset$, there exists $x \in A - B$. By the definition of $A - B$, $x \in A$ and $x \notin B$, hence there exists an element of A that is not in B. Therefore, $A \not\subseteq B$.

Case 2. Again we argue by contraposition. We assume $A \not\subseteq B$ and show $A - B \neq \emptyset$. Since $A \not\subseteq B$, there exists $x \in A$ such that $x \notin B$. Thus $x \in A - B$ and $A - B \neq \emptyset$.

The proof of part (i) is now complete. ∎

Proof of (vi)   First we use element-chasing to show $(A \cup B) - C \subseteq (A - C) \cup (B - C)$. Let $x \in (A \cup B) - C$. Then $x \in A \cup B$ and $x \notin C$. Therefore $(x \in A$ or $x \in B)$ and $x \notin C$, and hence either $(x \in A$ and $x \notin C)$ or $(x \in B$ and $x \notin C)$; i.e., $x \in A - C$ or $x \in B - C$, i.e., $x \in (A - C) \cup (B - C)$.

To show $(A - C) \cup (B - C) \subseteq (A \cup B) - C$, we let $x \in (A - C) \cup (B - C)$. Then either $x \in A - C$ or $x \in B - C$. In the first instance, $x \in A$ and $x \notin C$, hence $x \in A \cup B$ and $x \notin C$; in other words, $x \in (A \cup B) - C$. In the second case, $x \in B$ and $x \notin C$, which implies that $x \in A \cup B$ and $x \notin C$; again we conclude that $x \in (A \cup B) - C$. Therefore, $(A - C) \cup (B - C) \subseteq (A \cup B) - C$. ∎

The basic properties of the complement operation are recorded in the following theorem.

Theorem 3   *Let U be a set and let A and B be subsets of U.*

(i) $\emptyset^c = U$.
(ii) $(A^c)^c = A$.
(iii) $A \subseteq B$ *if and only if* $B^c \subseteq A^c$.
(iv)   *DeMorgan's Laws*
    (a) $(A \cap B)^c = A^c \cup B^c$.
    (b) $(A \cup B)^c = A^c \cap B^c$.

Note that $A^c = U - A$ is the complement of A in U.

Before considering the proofs, draw Venn diagrams illustrating each part of the theorem.

Proof   The proofs of (i) and (ii) are left as exercises.

To prove (iii), first we suppose $A \subseteq B$ and show $B^c \subseteq A^c$. This amounts to proving that if $x \in B^c$, then $x \in A^c$. Assume, to the contrary, that $x \notin A^c$. Then, by part (ii), $x \in (A^c)^c = A$; since $A \subseteq B$, $x \in B$. But also $x \in B^c$, implying that $x \notin B$ and forcing the absurd conclusion that $x \in B$ and $x \notin B$. This contradiction means that the assumption that $x \in A$ is incorrect. Thus $x \in A^c$, which completes the proof that $B^c \subseteq A^c$.

A similar argument establishes the converse. However, a slicker argument, using part (ii) and the implication that was proved in the previous paragraph, can also be given. We have just proved: $(*)$ If $A \subseteq B$, then $B^c \subseteq A^c$. Now we want to show that if $B^c \subseteq A^c$, then $A \subseteq B$.

By assumption $B^c \subseteq A^c$. From $(*)$ it follows that $(A^c)^c \subseteq (B^c)^c$. But by part (ii), $(A^c)^c = A$ and $(B^c)^c = B$. Thus $A \subseteq B$.

(Notice that the previous argument makes no reference to the definition of set inclusion: We do not show that $x \in A$ implies $x \in B$. Rather it uses previously established properties of set inclusion and set complement, and thus can be thought of as taking place at a level above the definitional level. This argument has several advantages—including brevity, elegance, and clarity—over a proof proceeding directly from the definition. Now back to the proof.)

We prove (iv)(a): $(A \cap B)^c = A^c \cup B^c$. Observe that once (iv)(a) is established, (iv)(b) can be derived as follows: By parts (iv)(a) and (ii),

$$(A^c \cap B^c)^c = (A^c)^c \cup (B^c)^c = A \cup B.$$

Thus, $(A \cup B)^c = ((A^c \cap B^c)^c)^c = A^c \cap B^c$ by part (ii).

To prove $(A \cap B)^c = A^c \cup B^c$, consider first the inclusion $(A \cap B)^c \subseteq A^c \cup B^c$. Let x be an arbitrary element of U such that $x \in (A \cap B)^c$. Then $x \notin A \cap B$; that is, it is not the case that x is an element of both A and B. Thus either $x \notin A$ or $x \notin B$; equivalently, either $x \in A^c$ or $x \in B^c$. Thus if $x \in (A \cap B)^c$, then $x \in A^c \cup B^c$.

Conversely, to show $A^c \cup B^c \subseteq (A \cap B)^c$, let x be an arbitrary element of U such that $x \in A^c \cup B^c$. We want to show $x \in (A \cap B)^c$, i.e., $x \notin A \cap B$. Since $x \in A^c \cup B^c$, $x \in A^c$ or $x \in B^c$; hence either $(x \in U$ and $x \notin A)$ or $(x \in U$ and $x \notin B)$. Thus $x \in U$ and $(x \notin A$ or $x \notin B)$. (Here we use the tautology $[P \wedge (Q \vee R)] \Leftrightarrow [(P \wedge Q) \vee (P \wedge R)]$.) It follows that $x \in U$ and it is not the case that $x \in A$ and $x \in B$. Therefore, $x \in U$ and $x \notin A \cap B$, i.e., $x \in (A \cap B)^c$. The proof of Theorem 3 part (iv) is now complete. ∎

THE POWER SET OF A SET

In order to represent intersection and union as operations, we had to consider the set of all subsets of a given set. This concept occurs everywhere in mathematics; the next definition supplies its official name and notation.

Definition 1   For any set A, the *power set of A*, written $P(A)$, is the set of all subsets of A. In other words, $P(A) = \{B \mid B \subseteq A\}$.

Example 1   Since $\mathbf{Z}, \mathbf{Q} \subseteq \mathbf{R}$, $\mathbf{Z}, \mathbf{Q} \in P(\mathbf{R})$.

Notice that the power set of A is a set each of whose elements is itself a set, namely a subset of A. Sets whose elements are sets can be confusing to work with, but perhaps a few examples can make the situation a little less forbidding.

Example 2   If A is any set, then $\emptyset \in P(A)$, $A \in P(A)$, and $\{a\} \in P(A)$ for each $a \in A$.

Example 3
   (i) $P(\emptyset) = \{\emptyset\}$.
   (ii) $P(\{1\}) = \{\emptyset, \{1\}\}$.
   (iii) $P(\{1,2\}) = \{\emptyset, \{1\}, \{2\}, \{1,2\}\}$.
   (iv) $P(\{1,2,3\}) = \{\emptyset, \{1\}, \{2\}, \{3\}, \{1,2\}, \{1,3\}, \{2,3\}, \{1,2,3\}\}$.

Let us look closely at the sets listed in Example 3. From those examples, a tight relationship seems to exist between the number of elements in a finite set and the number of subsets of the set. Let S_n denote the number of subsets in a set having n elements. (Thus, if n is the number of elements of A, then S_n is the number of elements of $P(A)$.) From Example 3 we can fill in the following table:

n	0	1	2	3
S_n	1	2	4	8

A pattern emerges, does it not? We *conjecture* that if A is a set with n elements then $P(A)$ is a set with 2^n elements.

Instead of searching for a proof of this conjecture immediately, let us glance at one more example. Let $A = \{1,2,3,4\}$. What is $P(A)$? To answer this we list the elements of $P(A)$ but let us do so in a systematic way. First, list the subsets of A that are also subsets of $\{1,2,3\}$, in other words those subsets of A that do not contain 4. There are 8 such subsets. What about the subsets of A that contain 4? Any such subset, B, contains 4 and possibly elements from $\{1,2,3\}$. Thus $B = \{4\} \cup C$ where C is a subset of $\{1,2,3\}$. Since there are exactly 8 such sets C, there are exactly 8 subsets of A that contain 4. Since any subset of A either contains 4 or does not contain 4, all subsets of A have been counted. Therefore, $P(A)$ has 16 elements.

Armed with this added evidence for our conjecture, we embark upon a proof. Certainly we would be wise to try to base our proof on an extension of the method used in finding the power set of $\{1,2,3,4\}$. Here is the theorem.

Theorem 4   *If A is a set with exactly n elements, then $P(A)$ has exactly 2^n elements.*

The proof of Theorem 4 will use the following lemma, which generalizes an observation made in our calculation of the power set of $\{1,2,3,4\}$. The lemma provides a more general result, since no restriction on the nature of the set A (finite or infinite) is placed.

Lemma 1   If $A = B \cup \{x\}$ where $x \notin B$, then $P(A) = P(B) \cup \{\{x\} \cup C \mid C \in P(B)\}$.

Proof   Let $D = P(B) \cup \{\{x\} \cup C \mid C \in P(B)\}$. We show $P(A) = D$ by checking that $P(A) \subseteq D$ and $D \subseteq P(A)$.

To show that $P(A) \subseteq D$ we take $A_1 \in P(A)$. Then $A_1 \subseteq A$. Consider two cases: (1) $x \notin A_1$, and (2) $x \in A_1$.

Case (1). Suppose $x \notin A_1$. Since $A_1 \subseteq A$, $A_1 \subseteq B$, hence $A_1 \in P(B)$.

Case (2). Suppose $x \in A_1$. Let $C = \{y \mid y \in A_1 \text{ and } y \neq x\}$. Then $C \subseteq B$, i.e., $C \in P(B)$ and $A_1 = \{x\} \cup C$. Therefore, in this case as well, $A_1 \in D$.

Thus $P(A) \subseteq D$.

The proof of the inclusion $D \subseteq P(A)$ is left as an exercise. With that, the lemma is proved. ∎

We are now ready to prove Theorem 4. Since the statement of the theorem contains a natural number parameter, a proof by mathematical induction is plausible.

Proof.   For $n = 1, 2, 3, \ldots$, let $S(n)$ be the statement: If A is a set with exactly n elements, then $P(A)$ has exactly 2^n elements.

We prove that $S(n)$ is true for all n by showing that
(i) $S(1)$ is true.
(ii) If $S(k)$ is true, then $S(k+1)$ is true.
(i) $S(1)$ is the statement: If A is a set with one element, then $P(A)$ has two elements. We have already verified this statement: If A has one element, then $A = \{a\}$ for some element a and $P(A) = \{\emptyset, A\}$ has two elements.
(ii) Suppose $S(k)$ is true and show $S(k+1)$ is true. Suppose A is a set with $k+1$ elements. Then we must show that $P(A)$ has 2^{k+1} elements. Let $A = \{a_1, \ldots, a_k, a_{k+1}\}$ and let $B = \{a_1, \ldots, a_k\}$. Then $A = B \cup \{a_{k+1}\}$. By Lemma 1, every subset of A either
   (a) does not contain a_{k+1} hence is a subset of B or
   (b) contains a_{k+1} and has the form $\{a_{k+1}\} \cup C$ where C is a subset of B (see Figure 4).

By inductive hypothesis, statement $S(k)$ is true. Thus any set with k elements has 2^k subsets. Since B has k elements, $P(B)$ has exactly 2^k elements. Therefore, there are 2^k subsets of A of type (1). In addition, as each subset of A of type (2) has the form $\{a_{k+1}\} \cup C$ where C is a subset of B, there are precisely 2^k subsets of A of type (2). But each subset of A is either of type (1) or type (2), and no subset of A can be of both types. Thus A contains $2^k + 2^k = 2^{k+1}$ subsets. ∎

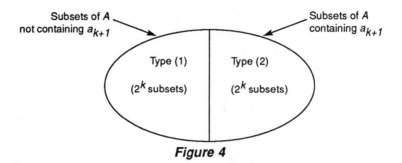

Figure 4

Remark   Notice that $P(B) \cap \{\{x\} \cup C \mid C \in P(B)\} = \emptyset$. In other words, each subset of A is either a subset of B or of the form $\{x\} \cup C$ where C is a subset of B, but no subset of A satisfies both these conditions.

Second Proof of Theorem 4   We outline another proof of Theorem 4, this one based on the binary sequence representation of a subset of a finite set.

Let $A = \{a_1, \ldots, a_n\}$ be a set having exactly n elements. The proof consists of two major steps:

(1)   There are precisely as many subsets of A as there are binary sequences of length n.

(2)   There are exactly 2^n binary sequences of length n.

The first step follows directly from the correspondence between subsets of A and binary sequences of length n defined in section 2. Simply observe that under this correspondence each subset of A is matched with exactly one binary sequence and each binary sequence corresponds to exactly one subset of A.

As you might imagine, the second step is proved by induction. Each binary sequence of length n ends either with 0 or 1. But how many such sequences end with 0? How many end with 1?   ∎

One might wonder why we bother to give a second proof of Theorem 4 (or any other theorem for that matter). The most immediate answer is that two proofs can only increase our understanding of the result. Each of the proofs given above is based on a specific way of picturing or representing the set $P(A)$. By using each representation to construct a proof, we are demonstrating the viability and power of that representation. By viewing a given result in different ways, we gain a clearer and richer perspective on the nature of that result. A closing remark about the name "power set" for $P(A)$. The source for the name can be seen in Theorem 4: *The number of elements in $P(A)$ is 2 to the power of the number of elements of A, when A is a finite set.* Thus, a property of $P(A)$ that holds when A is finite (namely, that $P(A)$ has 2^n elements if A has n elements) is used to suggest a name for $P(A)$ for arbitrary A. This property also motivates questions concerning the size of $P(A)$ when A is infinite. For example, when n is a large integer, say $n \geq 6$, 2^n is significantly larger than n.

a conjecture, draw a few pictures and calculate a few simple (e.g., finite) examples.

Theorem 5 *Let $A, B, C, and D$ be sets. Then*

(i) $A \times (B \cap C) = (A \times B) \cap (A \times C)$.

(ii) $A \times (B \cup C) = (A \times B) \cup (A \times C)$.

(iii) $(A \times B) \cap (C \times D) = (A \cap C) \times (B \cap D)$.

(iv) $(A \times B) \cup (C \times D) \subseteq (A \cup C) \times (B \cup D)$.

Proof of Theorem 5(i) We show that for all x, $x \in A \times (B \cap C)$ if and only if $x \in (A \times B) \cap (A \times C)$. From the definitions of intersection and Cartesian product,

$$x \in A \times (B \cap C)$$

if and only if $\quad x = (u, v)$ where $u \in A$ and $v \in B \cap C$,

if and only if $\quad x = (u, v)$ where $u \in A$, $v \in B$, and $v \in C$,

if and only if $\quad x = (u, v) \in A \times B$ and $x = (u, v) \in A \times C$,

if and only if $\quad x \in (A \times B) \cap (A \times C)$. ∎

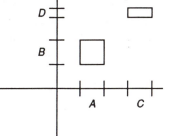

Figure 7 $(A \times B) \cup (C \times D)$.

Proof of Theorem 5(iv) To show that $(A \times B) \cup (C \times D) \subseteq (A \cup C) \times (B \cup D)$, we show that if $x \in (A \times B) \cup (C \times D)$, then $x \in (A \cup C) \times (B \cup D)$.

Since $x \in (A \times B) \cup (C \times D)$, either $x \in A \times B$ or $x \in C \times D$. If $x \in A \times B$, then $x = (a, b)$ where $a \in A$ and $b \in B$. Since $A \subseteq A \cup C$ and $B \subseteq B \cup D$, $x = (a, b) \in (A \cup C) \times (B \cup D)$.

On the other hand, if $x \in C \times D$, then $x = (c, d)$ where $c \in C \subseteq A \cup C$ and $d \in D \subseteq B \cup D$. Thus $x = (c, d) \in (A \cup C) \times (B \cup D)$. Therefore, if $x \in (A \times B) \cup (C \times D)$, then $x \in (A \cup C) \times (B \cup D)$.

We leave the proofs of (ii) and (iii) as exercises. ∎

Remark Let us draw a picture to illustrate Theorem 5(iv). We take A, B, C, and D to be intervals in $\mathbf{R}$. Then $(A \times B) \cup (C \times D)$ and $(A \cup C) \times (B \cup D)$ are pictured in Figures 7 and 8, respectively. These figures clearly illustrate the statement that $(A \times B) \cup (C \times D) \subseteq (A \cup C) \times (B \cup D)$.

(In fact, $2^n / n$ tends to ∞ as n tends to ∞). Thus $P(A)$ is significantly larger than A if A is a large finite set. Question: If A is infinite, is $P(A)$ larger than A? Before attempting to answer this question, we must clarify it. If A is infinite, then $P(A)$ is also infinite. (Why?) Thus, what does it mean to say that one infinite set is "larger" than another infinite set? How can the concept of size of a set be extended from finite sets to infinite sets? The answers to these questions lead directly to the concept of *cardinality* of a set, one of the most fascinating topics in all of mathematics. For more details, see references [2] or [4].

ORDERED PAIRS AND CARTESIAN PRODUCTS

Now we turn to yet another set operation, the Cartesian product. Our purpose is to define the Cartesian product of two sets and to establish some connection between Cartesian products and previously defined set operations.

The Cartesian product is familiar to students of pre-calculus and calculus in a special case, the Cartesian plane, $\mathbf{R}^2$. Let us recall the definition of $\mathbf{R}^2$ so as to use it as a motivating example. By definition,

$$\mathbf{R}^2 = \{(x, y) \mid x \in \mathbf{R} \text{ and } y \in \mathbf{R}\}.$$

The symbol (x, y) is called the ordered pair x and y. Thus to form $\mathbf{R}^2$, one "combines" pairs of elements of $\mathbf{R}$ in a certain fashion.

Geometrically, this combination is represented as a plane stretching endlessly in every direction with two specific perpendicular lines serving to provide a method of labeling points in the plane. Symbolically, the combination (x, y) of the elements x and y of $\mathbf{R}$ is regarded as a list with two entries, x being the first and y the second. The elements x and y are often called the *first* and *second coordinates* of (x, y), respectively.

The geometric representation of $\mathbf{R}^2$ is facilitated in part by the geometric representation of $\mathbf{R}$ as a straight line whose points are labelled by real numbers (Figure 5). Suppose we wish to extend the concept of Cartesian plane by considering, in place of $\mathbf{R}$, an arbitrary set A. Conceiving of this extension in geometric terms may be difficult unless a geometric representation of A is readily available. However, the symbolic description of $\mathbf{R}^2$ as a set of ordered pairs can be extended easily to an arbitrary set A : Simply take all ordered pairs (x, y) of elements of A where $x \in A$ is regarded as the first element and $y \in A$ is taken to be the second. Thus, for a given set A, we define

$$A^2 = \{(x, y) \mid x \in A \text{ and } y \in A\}.$$

Before setting this definition in concrete, let us take a closer look at the idea of ordered pair. In fact, no formal definition of the concept of ordered pair has been given; we have merely said that the symbol (x, y) is called the ordered pair x and y, where x is the first member of the pair and y

Before setting this definition in concrete, let us take a closer look at the idea of ordered pair. In fact, no formal definition of the concept of ordered pair has been given; we have merely said that the symbol (x, y) is called the ordered pair x and y, where x is the first member of the pair and y is the second. While the meaning of the term may be clear, the situation is unsatisfactory. Our understanding of the idea of ordered pair as "defined" above depends heavily on the written representation of it or on concepts that are difficult to define mathematically (at least at this stage of our development) such as "list with two entries." Thus, even though we may have a sound, intuitive feeling for the notion of ordered pair, we may find the idea elusive and difficult to grasp in situations where tight, rigorous reasoning is required. Thus, we demand a precise, mathematical definition of the term "ordered pair." To provide this definition, recall the dictum mentioned earlier in this section that any mathematical entity should be definable in terms of some set. With this in mind, we propose the following definition.

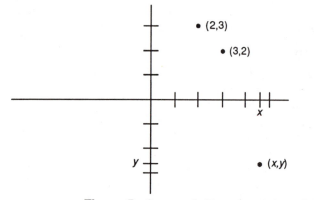

Figure 5   Geometric Representation of $\mathbf{R}^2$.

Definition 2   *Ordered Pair*

Let x and y be elements of set A and B, respectively. The *ordered pair x and y*, written (x, y), is defined to be the set

$$(x, y) = \{\{x\}, \{x, y\}\}.$$

Observe that (x, y) is a set with two elements (well, usually; see Exercise 14(a)) and each of these elements is itself a set. This definition allows us to distinguish between (x, y) and (y, x). For example, consider the pairs $(2, 3)$ and $(3, 2)$:

$$(2, 3) = \{\{2\}, \{2, 3\}\} \text{ and } (3, 2) = \{\{3\}, \{3, 2\}\}.$$

Definition 3   *Cartesian Product*

Let A and B be sets. The *Cartesian product of A and B*, written $A \times B$, is the set

$$A \times B = \{(a, b) \mid a \in A \text{ and } b \in B\}.$$

If $A = B$, then we write A^2 for $A \times B = A \times A$.

Example 4   Let $A = \{1, 2\}$ and $B = \{3, 4, 5\}$. Then

$$A \times B = \{(1, 3), (1, 4), (1, 5), (2, 3), (2, 4), (2, 5)\}$$

$$B \times A = \{(3, 1), (3, 2), (4, 1), (4, 2), (5, 1), (5, 2)\}, \text{ and}$$

$$A^2 = \{(1, 1), (1, 2), (2, 1), (2, 2)\}$$

Thus, in general, $A \times B \neq B \times A$. Question to investigate: Under wh conditions on A and B does $A \times B = B \times A$?

What about pictures for the Cartesian product? As mentioned earli the Cartesian plane $\mathbf{R}^2 = \mathbf{R} \times \mathbf{R}$ has a familiar geometric realizatio Because of our past experience with this set, we use $\mathbf{R}^2$ as a model examp for the Cartesian product of any two sets. Thus, whenever we need a pictu of the Cartesian product of two sets, we take the sets to be subsets of so that their Cartesian product is a subset of $\mathbf{R} \times \mathbf{R}$ (Figure 6).

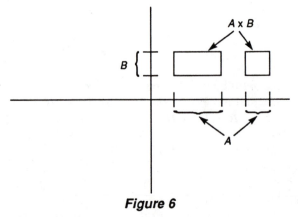

Figure 6

What is the relationship between the operation of Cartesian products and each of the set operations of union and intersection? For example, does $\times$ distribute over $\cap$ and $\cup$? Before reading the answer in the next theorem, you might wish to conjecture an answer. To help formulate

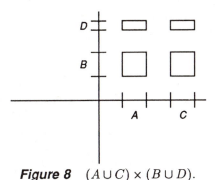

Figure 8 $(A \cup C) \times (B \cup D)$.

We close this section with a problem. Suppose A and B are finite sets containing, respectively, m and n elements where m and n are positive integers. Question: How many elements does the set $A \times B$ contain? Specifically, can we express the number of elements in $A \times B$ in terms of the integers m and n, the number of elements in A and B, respectively?

To investigate the question, let us take a specific example: Let $A = \{1, 2\}$ and $B = \{3, 4, 5\}$. Then $A \times B = \{(1, 3), (2, 3), (1, 4), (2, 4), (1, 5), (2, 5)\}$. Thus $A \times B$ contains six elements while A and B contain two and three elements, respectively. Now let $A = \{0, 1, 2\}$ and $B = \{3, 4, 5\}$. Then $A \times B$ contains nine elements. The examples chosen above have the advantage of being simple enough to allow for easy calculation, yet complex enough to suggest, perhaps, an answer to our question. (If these examples do not suggest an answer, then try more examples.) On the basis of these examples, one might conjecture the following statement:

If A and B are sets containing, respectively, m and n elements where m and n are positive integers, then $A \times B$ contains mn elements.

Before trying to prove the result, let us consider a rather special case. Suppose $A = \{a\}$ contains just one element and $B = \{b_1, \ldots, b_n\}$ is any finite set with n elements. Then $A \times B = \{(a, b_1), \ldots, (a, b_n)\}$ contains exactly $n = 1 \cdot n$ elements. Thus our conjecture is confirmed in this case. Similarly, it is easy to check that if B contains just one element and A contains m elements, then $A \times B$ contains $m = 1 \cdot m$ elements.

Now let us try to prove the conjecture. Which proof method should we consider? An indirect proof does not seem appropriate. Since integer variables (m and n are integers) are involved, perhaps a proof by mathematical induction is worth trying. The trouble is that there are two integer variables. We can handle this obstacle by quantifying matters appropriately. For each positive integer n, let $S(n)$ be the statement:

$S(n)$: If A is a set containing exactly m elements where m is a positive integer and B is a set containing n elements, then $A \times B$ contains $m \cdot n$ elements.

Proof by mathematical induction

Basis step: If $n = 1$, then $B = \{b\}$. Suppose A contains m elements where $m \in \mathbf{Z}^+$. Then $A = \{a_1, \ldots, a_m\}$. Thus $A \times B = \{(a_1, b), \ldots, (a_m, b)\}$ and $A \times B$ contains $m = m \cdot 1 = m \cdot n$ elements. Therefore, $S(1)$ is true.

Inductive step: We show that if $S(n)$ is true, then $S(n+1)$ is true. Thus we must prove that if A contains m elements and B contains $n+1$ elements, then $A \times B$ contains $m \cdot (n+1)$ elements.

Since A has m elements, $A = \{a_1, \ldots, a_m\}$. Since B contains $n+1$ elements, $B = \{b_1, \ldots, b_n, b_{n+1}\}$. Keep in mind that at some point we want to use the inductive hypothesis, i.e., we want to use the assumption that $S(n)$ is true. We can do this by referring to Theorem 5(ii): Since $B = \{b_1, \ldots, b_m, b_{m+1}\} = \{b_1, \ldots, b_n\} \cup \{b_{m+1}\}$,

$$A \times B = A \times (\{b_1, \ldots, b_n\} \cup \{b_{n+1}\})$$
$$= (A \times \{b_1, \ldots, b_n\}) \cup (A \times \{b_{n+1}\})$$

Now the set $A \times \{b_1, \ldots, b_n\}$ contains $m \cdot n$ elements, since A contains m elements and $\{b_1, \ldots, b_n\}$ contains n elements, and since $S(n)$ is true. Since $S(1)$ is true, $A \times \{b_{n+1}\}$ contains m elements. In addition, the sets $A \times \{b_1, \ldots, b_n\}$ and $A \times \{b_{n+1}\}$ possess no common elements, i.e., $A \times (\{b_1, \ldots, b_n\}) \cap (A \times \{b_{n+1}\}) = \emptyset$. Why? It follows from Exercise 20 at the end of this section that $A \times B = A \times (\{b_1, \ldots, b_n\}) \cup (A \times \{b_{n+1}\})$ contains $m \cdot n + m = m \cdot (n+1)$ elements. Thus, $S(n+1)$ is true if $S(n)$ is true.

EXERCISES §10

1. Let $U = \{1, 2, 3, 4, 5, 6\}$, $A = \{1, 2, 3\}$, $B = \{2, 3, 4\}$, and $C = \{4, 5, 6\}$. Find each of the following sets.

   (a) $A \cup B$ (e) $B \cap C$ (i) $A - C$
   (b) $A \cap B$ (f) $A \cap (B \cup C)$ (j) $P(A) \cap P(B)$
   (c) A^c (g) $(A \cap B) \cup (A \cap C)$
   (d) B^c (h) $A \cap C^c$

2. Let A and B be arbitrary sets.
   (a) Prove: $A \cup A = A$ and $A \cap A = A$.
   (b) Prove: $A \cup B = B \cup A$ and $A \cap B = B \cap A$.

3. (a) Prove: $A \subseteq B$ if and only if $A \cup B = B$.
   (b) Prove: $A \subseteq B$ if and only if $A \cap B = A$.
   (c) Prove: If $A \subseteq B$ and $A \subseteq C$, then $A \subseteq B \cap C$.
   (d) Prove: If $A \subseteq B$ and $C \subseteq D$, then $A \cap C \subseteq B \cap D$ and $A \cup C \subseteq B \cup D$.

4. Let U be a set and A, B, and C be subsets of U.
   (a) Prove: $A - B = A \cap B^c$ and $A - B = B^c - A^c$.
   (b) Prove: $A - (A - B) = A \cap B$.
   (c) Prove: $(A - B)^c = A^c \cup B$.

(d) Prove: $(A \cup B) - (A \cap B) = (A - B) \cup (B - A)$.

5. Prove Theorem 2, parts (ii)–(v).

6. Find $P(\{1\})$ and $P(P(\{1\}))$.

7. Prove: If $A \cap B = \emptyset$, then $P(A) \cap P(B) = \{\emptyset\}$. Is the converse true?

8. (a) Prove: $P(A \cap B) = P(A) \cap P(B)$. Express this result in a sentence that contains no mathematical symbols.

   (b) State and prove a result concerning the power set of the union of two sets.

9. Suppose A is a set with n elements where n is a positive integer.

   (a) How many subsets of A contain exactly one element?

   (b) How many subsets of A contain exactly $n - 1$ elements? (Hint: consider the binary representation of subsets.)

10. (a) Find $A \times B$ if $A = \{1, 2, 3\}$ and $B = \{3, 4, 5\}$.

    (b) Find $\{a, b\} \times \{a, b\}$.

    (c) Find $\{(a, b)\} \times \{a, b\}$.

11. Draw a sketch of each of the following subsets of $\mathbf{R}^2$:

    (a) $[0, 1] \times [1, 2]$ (Recall $[a, b] = x \in \mathbf{R} \mid a \le x \le b$).

    (b) $([0, 1] \cup \{2\}) \times [1, 2]$.

    (c) $([0, 1] \cup \{2\}) \times ([1, 2] \cup \{3\})$.

12. Prove: $(a, b) = (c, d)$ if and only if $a = c$ and $b = d$.

13. (a) Show: $(a, a) = \{\{a\}\}$.

    (b) Show: $\{a\} \times \{a\} = \{\{\{a\}\}\}$.

14. True or false: $(a, b) \cap (b, a) = \emptyset$.

15. Fill in the blank: $A \times B = \emptyset$ if and only if _____.

16. (a) Prove: If $a, b \in A$, then $(a, b) \in P(P(A))$.

    (b) Prove: If $a \in A$ and $b \in B$, then $(a, b) \in P(P(A \cup B))$.

17. Let A and B be sets. State a necessary and sufficient condition for $A \times B$ to equal $B \times A$.

18. Prove or disprove and salvage: If $A, B, C,$ and D are all nonempty sets, then $(A \times B) \cup (C \times D) \subset (A \cup C) \times (B \cup D)$.

19. Prove or disprove and salvage: If $A, B,$ and C are sets with $A \times B = A \times C$, then $B = C$.

20. Prove by induction that if A and B are finite sets, respectively containing m and n elements where m and n are positive integers and $A \cap B = \emptyset$, then $A \cup B$ contains $m + n$ elements.

We continue with our discussion of set theory. We concentrate on a fundamental theme, that of *relation*, and several important variations on this theme, namely those of order relations, equivalence relations, and functions. To varying degrees these ideas are familiar to most readers: functions, of course, have been with us since high school where linear and quadratic equations were analyzed thoroughly. Ideas such as "less than" in algebra and "congruence" in geometry are examples of order relations and equivalence relations, respectively.

The idea of a relation arises out of the consideration of connections among elements of a given set or among elements of one set and elements of another set. For example, two people, A and B, living today might be connected by the fact that they are siblings. Obviously, some pairs of people are connected in this way; most are not. As for some mathematical examples, two real numbers, x and y, might be connected by the fact that $y = e^x$, or x and y might be related by the fact that x is less than or equal to y. A precise mathematical idea, called a *relation*, hides behind these illustrations. We will propose a formal definition after considering the last example in more detail.

Recall that if x and y are real numbers, then x is *less than or equal to* y, written $x \leq y$, if $y - x$ is nonnegative. Thus, given a pair of real numbers, in fact an ordered pair of real numbers (x, y), to determine if $x \leq y$, we simply check if $y - x$ is nonnegative; equivalently, we see whether y can be obtained by adding some nonnegative number to x. If this is not the case, then we write $x \nleq y$. For example, $e \leq \pi$, $\pi^2 \leq 10$, but $\pi \nleq e$. Hence, the statement "x is less than or equal to y" describes a property of the ordered pair (x, y). We can then use this property to define a set. Consider the subset, call it $R_\leq$, of $\mathbf{R}^2$ defined as follows:

$$R_\leq = \{(x, y) \in \mathbf{R}^2 \mid x \leq y\}$$
$$= \{(x, y) \mid y - x \text{ is nonnegative}\}.$$

The set $R_\leq$ determines the relationship between x and y relative to the property "less than or equal"; for $x \leq y$ if and only if $(x, y) \in R_\leq$. Generalizing from this example, we introduce the following simple yet important concept.

Definition 1   *Relation*

A *relation on a set* A is a subset, R, of $A^2 = A \times A$.

Being a subset of $A \times A$, a relation is specified in the ways that subsets of a given set are generally defined. For instance, if A is a finite set (preferably a small finite set), then a relation on A can be defined simply by listing its elements. Of course rarely is this method practical or interesting. Usually to define a relation, we provide a statement (more precisely, a predicate) that singles out a collection of elements of $A \times A$ for membership in the relation. For any relation R, one writes, of course, $(x, y) \in R$ if the pair (x, y) satisfies the defining property of R. In practice, we usually drop this notation in favor of a more natural way of writing things. Again with the example of $\leq$ in mind, we write $x \, R \, y$ if $(x, y) \in R$. Let us look at several examples.

Example 1   For $x, y \in \mathbf{R}$, define $x < y$ to mean $y - x$ is positive. Thus $x < y$ if and only if $x \leq y$ and $x \neq y$.

Example 2   When first encountering a new idea, it always helps to see "extreme" examples of it, even if these examples happen to be trivial and uninteresting. For any set A, there are two extreme examples of relations on A, namely $\emptyset$ and A^2 (and, indeed, these examples are not very interesting). In the first case, no element is related to any other, and in the second case, each element is related to all others. These examples are "extreme" in the sense that if R is any relation on A, then $\emptyset \subseteq R \subseteq A \times A$. They are "trivial" in the sense that for any set A, $\emptyset$ and $A \times A$ are relations on A.

Example 3   Let $A = \{1, 2\}$. Here are several relations on A: $R_1 = \{(1, 1), (1, 2)\}$, $R_2 = \{(1, 1), (2, 2), (1, 2)\}$, $R_3 = \{(1, 2)\}$, $R_4 = \{(2, 1)\}$, and $R_5 = \{(1, 2), (2, 1)\}$. Since $A \times A$ has 4 elements, $P(A \times A)$ has $2^4 = 16$ subsets, and thus there exist 16 relations on $A = \{1, 2\}$.

Example 4   Here is another simple, general (and familiar) relation. For any set A let

$$I_A = \{(x, y) \mid x = y\}.$$

Although I_A merely provides a formal way of expressing equality, it usually goes by the name *identity relation*.

Example 5   Let $\mathbf{Z}^+$ denote the set of positive integers. (Thus $\mathbf{Z}^+ = \mathbf{N} - \{0\}$.) For $a, b \in \mathbf{Z}^+$, one says that a *divides* b, written $a \mid b$, if there exists $c \in \mathbf{Z}^+$ such that $ac = b$. Write $a \nmid b$ if a does not divide b. Examples: $2 \mid 8$, $7 \mid 91$, and $3 \nmid 89$. Thus, the common notion of divisibility of positive integers yields an example of a relation. This relation can be extended to the set $\mathbf{Z}$ of all integers in an obvious way: For $a, b \in \mathbf{Z}$, $a \mid b$ means that there exists $c \in \mathbf{Z}$ such that $a \cdot c = b$. If

$a \mid b$, then we also say b is *divisible* by a, b is a *multiple* of a, a is a *factor* of b, or a is a *divisor* of b.

Example 6   Let U be a set.

(i) Containment defines a relation, $R_{\subseteq}$, on $P(U)$: $(A, B) \in R_{\subseteq}$ if and only if $A \subseteq B$.

(ii) For $x, y \in P(U)$, define xRy to mean $x \cap y \neq \emptyset$. R is a relation on $P(U)$.

Example 7   Let us recall another familiar example. Let $\mathcal{F}$ be the set of fractions of integers; 1/2, 2/3, 8/9, -13/11, etc. Define $a/b \cong c/d$ if $ad = bc$. Then $\cong$ is a relation on $\mathcal{F}$. Moreover, $\cong$ is precisely the relation that we employ when we assert that two ways of writing a number as a fraction are equivalent. For instance, when we say that $1/2 = 2/4$, we are merely saying that the pair $(1/4, 2/4)$ is a member of the relation $\cong$.

Properties of Relations

We continue our study of relations by isolating various properties and studying those relations that have certain combinations of these properties. Several important relations happen to possess many of these properties. Actually, this fact is at least partly responsible for the formulation of these concepts. Later we will try to get a different view of properties of relations by interpreting them geometrically.

Definition 2   *Reflexive Relation*

A relation R on a set A is *reflexive* if, for all $x \in A$, $x R x$.

Definition 3   *Symmetric Relation*

A relation R on a set A is *symmetric* if, for all $x, y \in A$, $x R y$ implies $y R x$.

Definition 4   *Antisymmetric Relation*

A relation R on a set A is *antisymmetric* if, for all $x, y \in A$, $x R y$ and $y R x$ implies $x = y$.

Definition 5   *Transitive Relation*

A relation R on a set A is *transitive* if, for all $x, y, z \in A$, $x R y$ and $y R z$ implies $x R z$.

Example 8   Consider the inequality relations on **R** (or, if you like, on **Q**, **Z**, or **N**).

(i) $\leq$ is reflexive, antisymmetric, and transitive. The relation $\geq$ has the same properties.

(ii) The relation $<$ is transitive.

Example 9   Let U be a set and consider the inclusion relations on $P(U)$.

(i) $\subseteq$ is reflexive, antisymmetric, and transitive. The relation $\supseteq$ possesses the same properties.

(ii) The relation $\subset$ is transitive.

Example 10   Let $A = \{a, b, c\}$.

(i) The relation $R = \{(a,a), (a,b), (b,a), (b,c)\}$ is neither reflexive, symmetric, antisymmetric, nor transitive. Nor, for that matter, is R very interesting; however, it does illustrate the fact that simple examples of relations not having (or having) various properties can be easily constructed.

(ii) Let $R = \{(a,a), (b,b), (a,b)(c,c)\}$. Then R is a reflexive, antisymmetric, and transitive relation on A. Notice that R is not symmetric.

Example 11   The divisibility relation $|$ on **N** is reflexive, antisymmetric, and transitive. (Recall that for $a, b \in \mathbf{Z}^{+}, a \mid b$ if there exists $c \in \mathbf{N}$ such that $a \cdot c = b$.)   For instance, to prove transitivity, suppose $a \mid b$ and $b \mid c$ and show $a \mid c$. If $a \mid b$ and $b \mid c$, then there exist $x, y \in \mathbf{Z}$ such that $b = a \cdot x$ and $c = b \cdot y$. Therefore, $c = b \cdot y = (a \cdot x) \cdot y = a \cdot (x \cdot y)$, which implies that $a \mid c$. On **Z**, however, this relation is merely reflexive and transitive. Since $2 \mid (-2)$, $(-2) \mid 2$, and $-2 \neq 2$, the relation $|$ is not antisymmetric on **Z**.

As we saw in Examples 8 and 9, the relations $\leq$ on **R** and $\subseteq$ on $P(U)$ are reflexive, antisymmetric, and transitive. At the same time these relations provide ways of comparing the elements of their respective sets. For instance, if $A, B \in P(U)$ and $A \subseteq B$, then we can think of A as being "smaller than" B. We now extend the notions of inequality and ordering from these examples to arbitrary sets by isolating the basic properties of these relations. Unfortunately, the terminology is not completely standard; we will use what is probably the most common.

Definition 6   *Partial Ordering*

A relation R on a set A is a *partial ordering* if R is a reflexive, antisymmetric, and transitive relation. We write (A, R) to denote the set A and the partial ordering R.

As noted earlier, the relations $\leq$ and $\geq$ on **R** and $\subseteq$ and $\supseteq$ on $P(U)$ are all partial orderings. By contrast none of the relations $<$, $>$, $\subset$, or $\supset$ is a partial ordering.

Example 12   (i) The relation $|$ on $\mathbf{N^+}$ is a partial ordering.

(a) Reflexivity: For any $a \in \mathbf{N^+}$, $a \mid a$ since $a \cdot 1 = a$.

(b) Antisymmetry: If $a \mid b$ and $b \mid a$, then $b = ac$ and $a = bd$ for some $c, d \in \mathbf{N^+}$; then $b = ac = b(dc)$, which means that $dc = 1$, which in turn implies that $c = d = 1$, and $b = a$.

(c) Transitivity: The transitivity of $|$ on $\mathbf{N}$ was established in Example 11. It follows easily that $|$ is transitive on $\mathbf{N^+}$.

(ii) For any $n \in \mathbf{N^+}$, let $D_n = \{d \in \mathbf{N} \mid d \text{ divides } n\}$. Then the relation $|$ defines a partial ordering on D_n.

We have introduced partial orderings since they are natural and important examples of relations. We will not, however, pursue partial orderings and other order relations further.

Another important type of relation is an equivalence relation. The concept of equivalence relations is significant because it provides a natural way of generalizing the notion of equality. For the time being we present the definition and note a few examples.

Definition 7   *Equivalence Relation*

A relation R on a set A is an *equivalence relation* if R is reflexive, symmetric, and transitive.

Example 13   (i) Let A be an nonempty set. Then the identity relation I_A is an equivalence relation on A. Note that $\emptyset$ is not an equivalence relation on A since $\emptyset$ is not reflexive.

(ii) The relation $\cong$ on the set $\mathcal{F}$ of fractions of integers defined in Example 7 is an equivalence relation on $\mathcal{F}$.

In addition to partial orderings and equivalence relations, there is another important type of relation that plays a major role in mathematics. This class of relations is actually quite familiar, having been part of our mathematical diet since high school.

Definition 8   *Function*

For any set A, a *function on* A is a relation, f, on A with the property that for all $a, b, c \in A$ if $(a, b) \in f$ and $(a, c) \in f$, then $b = c$.

What does this condition say in ordinary mathematical English? It asserts that for each $a \in A$ there is at most one $b \in A$ such that $(a, b) \in f$, or $a f b$. Note that for a given $a \in A$ there may exist no element $b \in A$ for which $(a, b) \in f$, but if there is one such b, then there is exactly one. Since this element b is unambiguously determined by a, we can and should express this dependence notationally, and we do so by following the usual custom and writing $b = f(a)$. Note that this notation is both efficient and clear; for when we define $f(a)$, we are saying that $(a, f(a))$ is the unique element of f with first coordinate a.

In order to define a function, one must describe, for each $a \in A$ or for each a in some given subset of A, the element $f(a) \in A$ that corresponds to a. When $A = \mathbf{R}$, this correspondence can be based on concepts and constructions from arithmetic, geometry, or calculus. Here are some illustrations.

Example 14   In each of the following examples, f will be a function on $\mathbf{R}$.
  (i) $f(x) = x$ for $x \in \mathbf{R}$. (Note that $f = I_{\mathbf{R}}$.)
  (ii) $f(x) = x^2 + 1$ for $x \in \mathbf{R}$.
  (iii) $f(x) = \sin(x)$ for $x \in \mathbf{R}$.
  (iv) $f(x) = e^x$ for $x \in \mathbf{R}$.
  (v) $f(x) = \int_0^x (t^2 + 1)\, dt$ for $x \in \mathbf{R}$.
  (vi) $f(x) = \sum_{n=1}^{\infty} (x^n/n^n)$ for $x \in \mathbf{R}$.
  (vii) $f(x) = \sqrt{x}$ for $x \geq 0$.

Perhaps one aspect of our development of relations has seemed somewhat restrictive. By our definition, a relation on a set A is a subset of $A \times A$. As a general example of a relation on A, we introduced the concept of a function on A. But, as was apparent in calculus, one often works with functions, f, "from a set A to a set B" where A and B are possibly distinct sets. In this case, f assigns to each $x \in A$ a unique element $f(x) \in B$, hence f can be described as the subset $\{(x, f(x)) \in A \times B \mid x \in A\}$ of $A \times B$. The next definition extends this idea to relations.

Definition 9   *Relations and Functions*

Let A and B be sets.
  (i)   A *relation* from A to B is a subset R of $A \times B$.
  (ii)   A *function* from A to B is a relation f from A to B with the property that if $(x, y) \in f$ and $(x, z) \in f$, then $y = z$. For a function f from A to B, we write $y = f(x)$ if $(x, y) \in f$.
  (iii)   If f is a function from A to B with the property that for each $a \in A$ there exists $b \in B$ such that $(a, b) \in f$, then we write $f : A \rightarrow B$.

Warning: When we write $f : A \rightarrow B$, we mean that for each $a \in A$, there exists $b \in B$ such that $(a, b) \in f$. When we say that f is a function from A to B, we allow for the possibility that there exists $a \in A$ such that for all $b \in B$, $(a, b) \notin f$.

Example 15   Here is an example of a relation from outside mathematics. Let F be the set of American females and M the set of American males. Define

$$R = \{(x, y) \in F \times M \mid x \text{ is the mother of } y\}.$$

Then R is a relation from F to M.

REPRESENTATION OF RELATIONS

Is there a handy geometric way of representing relations? In posing this question, we have several goals in mind. First, we seek both concrete pictures of specific relations and general ways of viewing typical or model relations. Second, once we have pictures of relations, we can interpret various properties of relations visually, thereby obtaining a deeper understanding of these properties. For example, what does it mean pictorially for a relation to be symmetric? Presently, we will impose other conditions on relations, and in several cases will be able to describe these conditions in geometric terms.

By now, you might have guessed what our geometric interpretation of a relation will be. As we did when we constructed geometric realizations of Cartesian products, let us assume that A is a subset on $\mathbf{R}$. Then any relation on A, being a subset of A^2, is also a subset of $\mathbf{R}^2$. Thus we can sketch the set of points $\{(x, y) \in \mathbf{R}^2 \mid (x, y) \in R\}$; this picture is called the *graph of R*. Figure 1 shows the graph of a typical, garden-variety relation R on $\mathbf{R}$.

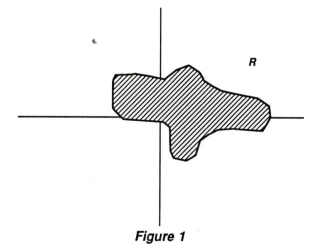

Figure 1

Conversely, any geometric figure or region in the Cartesian plane represents a relation on $\mathbf{R}$. Thus, the routine curves and figures of calculus and analytic geometry, which include parabolas, hyperbolas, ellipses, and graphs of functions, all yield relations on $\mathbf{R}$ (Figure 2).

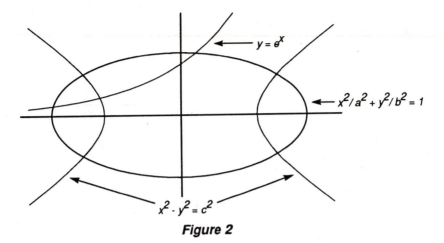

Figure 2

Reflexivity and symmetry have nice geometric interpretations. Consider a relation R on $\mathbf{R}$. If R is reflexive, then $x\,R\,x$ for all x in $\mathbf{R}$; hence the graph of R contains the diagonal line $y = x$ in $\mathbf{R}^2$ (Figure 3).

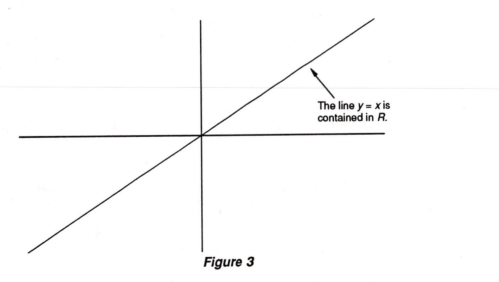

The line $y = x$ is contained in R.

Figure 3

If R is symmetric, then for all $x, y \in \mathbf{R}$, $x\,R\,y$ implies $y\,R\,x$. Now, for any $x, y \in \mathbf{R}$, the line in $\mathbf{R}^2$ joining (x, y) to (y, x) is perpendicular to and bisected by the line $y = x$ (Figure 4). Thus the graph of any symmetric relation, R, is symmetric in a geometric sense about the line $y = x$. Take the mirror-image of the graph of R about the line $y = x$; this image coincides with the graph of R precisely when R is symmetric (Figure 5).

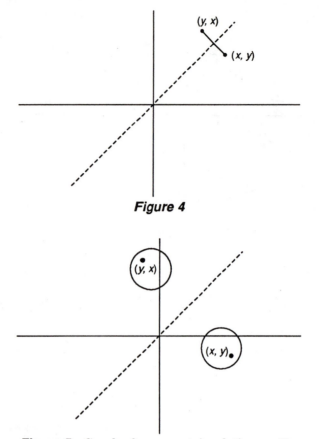

Figure 4

Figure 5   Graph of a symmetric relation on **R** .

Finally, a crude but effective way of representing a relation R from A to B (and in particular a function from A to B) is the following picture, which we call the *sketch* of R (Figure 6). Draw two closed figures to represent A and B. For $a \in A$ and $b \in B$, if $(a,b) \in R$, then label points in A and B as a and b, respectively, and join these points by a directed segment, i.e., a segment with an arrow at b.

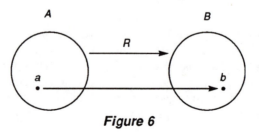

Figure 6

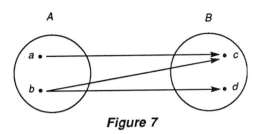

Figure 7

For example, let $A = \{a, b\}$ and $B = \{c, d\}$ and let $R = \{(a, c),$ $(b, c), (b, d)\}$. Then R is represented as depicted in Figure 7.

EXERCISES §11

1. How are the subsets $R_\leq$, $R_<$, $R_\geq$, and $R_>$ of $\mathbf{R}^2$ related? Sketch the graphs of these relations.

2. Give several examples of relations on $\{a, b, c\}$.

3. (a) How many relations on $\{1, 2\}$ are reflexive?
   (b) How many are symmetric?
   (c) How many are both reflexive and symmetric?
   (d) How many are neither reflexive nor symmetric?

4. (a) How many relations exist on the set $\{a, b, c\}$?
   (b) Suppose A is a finite set with n elements. How many relations on A are there?

5. On a single set of your choice, give examples of relations that possess exactly one and exactly two of the following three properties— reflexivity, symmetry, and transitivity. You should give six examples.

6. Can a relation be both symmetric and antisymmetric? Explain.

7. (a) Prove: On a set A, $\emptyset$ is a symmetric and transitive relation.
   (b) Prove: The identity relation on any set is a partial ordering.
   (c) Prove: On a set A having at least two elements, $A \times A$ is not a partial ordering.

8. Characterize graphs of functions on $\mathbf{R}$ among graphs of relations on $\mathbf{R}$.

9. Let R and S be relations on a set A. In each case, prove or disprove the given statement.
   (a) If R and S are reflexive, then $R \cap S$ is reflexive.
   (b) If R and S are reflexive, then $R \cup S$ is reflexive.
   (c) If R and S are symmetric, then $R \cap S$ is symmetric.
   (d) If R and S are symmetric, then $R \cup S$ is symmetric.
   (e) If R and S are transitive, then $R \cap S$ is transitive.
   (f) If R and S are transitive, then $R \cup S$ is transitive.

Section 12
FUNCTIONS

By definition, a function is actually a special kind of relation. The purpose of this section is to record some general properties of functions and to discuss various ways of representing functions. First, we recall the basic definition.

Definition 1 *Function*

Let A and B be sets. A *function* from A to B is a relation, f, from A to B such that if for $a \in A$ and $b, c \in B$, $(a, b) \in f$ and $(a, c) \in f$, then $b = c$. If $(a, b) \in f$, then we write $b = f(a)$. A function from A to B is also called a *mapping* from A to B.

Several technical terms and concepts are associated with the function concept. These names and ideas provide the vocabulary in which discussions involving functions take place. Most of the following ideas are familiar from calculus, although some of the labels for them might be new.

Definition 2 *Domain and Range*

If f is a function from A to B, then
  (i) the *domain of f*, written Dom (f), is the set: Dom $(f) = \{a \in A \mid$ there exists $b \in B$ such that $b = f(a)\}$.
  (ii) the *range of f*, written Ran(f), is the set: Ran$(f) = \{b \in B \mid$ there exists $a \in A$ such that $b = f(a)\}$.

Usually, when we consider a function from A to B, we assume that $A = \text{Dom}(f)$. In this case we write $f : A \to B$ to denote the function f. Note, however, that when we use this notation, we are not assuming that Ran$(f) = B$, merely that Ran$(f) \subseteq B$. (Note: Some authors call B the *codomain* of f.)

The symbol $f : A \to B$ is suggestive of the definition of a function that is used in less formal mathematics: A function from A to B is a rule or correspondence, f, that assigns to each $a \in A$ a unique element $f(a) \in B$. The "correspondence" definition of function has a special virtue: It suggests that a function is a dynamic, as opposed to a static, entity that transforms elements from one set into elements of another set. The view of a function as the embodiment of an active process assists in the visualization of functions and reflects the way in which functions actually arise in many applications. We next consider a special set of functions called sequences. Sequences appear throughout mathematics and in a wide range of applications of mathematics. Sequences play a particularly important role in discrete mathematics, the discipline that is concerned with properties of finite and denumerable sets (a certain class of infinite sets). The mathematical problems that arise in computer science are often problems in discrete mathematics and lead in many cases to questions involving sequences. Often in these situations, the sequences that occur are approximated by continuous functions, hence are studied using techniques of calculus.

Definition 3   *Sequences*

Let A be a set. A *sequence with values in A* is a function $s : \mathbf{N} \to A$. For $n \in \mathbf{N}$ one often writes s_n in place of $s(n)$, and denotes the sequence s by $\{s_n \mid n \in \mathbf{N}\}$.

Example 1   (i) Define $p : \mathbf{N} \to \mathbf{N}$ by $p(n) = p_n = 2^n$. Thus, p is the sequential analog of the base 2 exponential function $f(x) = 2^x$ whose domain is all of $\mathbf{R}$.

(ii) Define $f : \mathbf{N} \to \mathbf{N}$ by $f_0 = 1 = f_1$ and for $n \geq 2$, $f_n = f_{n-1} + f_{n-2}$. The sequence $\{f_n \mid n \in \mathbf{N}\}$ is called the *Fibonacci sequence* and is one of the most famous sequences in mathematics. (The interested reader might wish to consult the *Fibonacci Quarterly*, a research journal that publishes papers that focus on the Fibonacci sequence and related matters.)

(iii) Define the *factorial* function fact : $\mathbf{N} \to \mathbf{N}$ by $fact(0) = 1$ and for $n \geq 1$ $fact(n) = n \cdot fact(n-1)$. Thus, for example, $fact(3) = 3 \cdot fact(2) = 3 \cdot 2 \cdot fact(1) = 3 \cdot 2 \cdot 1 \cdot fact(0) = 3 \cdot 2 \cdot 1 \cdot 1 = 6$. It is customary to write $fact(n) = n!$ and to read $n!$ as "n factorial." The following table gives the first few values of the factorial sequence.

n	0	1	2	3	4	5	6	7	8	9	10
n!	1	1	2	6	24	120	720	5040	40320	362880	3628800

PROPERTIES OF FUNCTIONS

As we have just mentioned, a function is often regarded as a means of corresponding to each element of a certain set an element in another set. Several questions are suggested by this point of view. For example, if $f : A \to B$, then does each element of B have an element of A corresponding to it under f? If $f : A \to B$, then do two distinct elements of A correspond to the same element of B under f? The next definition provides a vocabulary for phrasing these questions.

Definition 4   *Injective, Surjective, and Bijective Functions*

Let $f : A \to B$ be a function.

(i) If, for every $x, y \in A$ such that $x \neq y$, $f(x) \neq f(y)$, then f is called *injective* or *one-to-one*.

(ii) If $\text{Ran}(f) = B$, then f is called *surjective* or *onto*.

(iii) If f is both injective and surjective, i.e. both one-to-one and onto, then f is called a *bijective* function or a *one-to-one correspondence*.

As it happens, all the terms presented in this definition are used very frequently. Thus, we should be conversant with all of them.

Example 2   Let $A = \{1, 2, 3\}$ and $B = \{4, 5\}$ and let $f : A \to B$ be defined by $f(1) = f(3) = 5$ and $f(2) = 4$. Then f is surjective but not injective. Question: Does there exist any injective function $g : A \to B$?

Let $f : A \to B$. How does one prove that f is surjective? The most frequently used strategy is to take an arbitrary $b \in B$; then, using the definition of f, one finds $a \in A$ such that $f(a) = b$. To show that f is one-to-one, two logically equivalent approaches are used: (i) Take $x, y \in A$ such that $x \neq y$ and show $f(x) \neq f(y)$, or (ii) suppose $f(x) = f(y)$ where $x, y \in A$ and show $x = y$.

Intuitively, we think of a function $f : A \to B$ as providing a way of carrying the set A to the set B. When f is injective, distinct elements of A are taken to distinct elements of B, and hence an injective function "preserves" the structure of the set A. Specifically, the subset of B

$$f(A) = \{y \in B \mid y = f(x) \text{ for some } x \in A\}$$

is a "copy" of A in the sense that f is a one-to-one correspondence between A and $f(A)$ (Figure 1). Injective functions need not be surjective; thus, a portion of B might be missed by f. When f is surjective, each $y \in B$ has the form $y = f(x)$ for some $x \in A$, and hence a surjective function $f : A \to B$ "covers" B with elements of A (Figure 2). A surjective function need not be injective, thus portions of A might "collapse" under f. Finally, a bijective function f establishes a matching between the sets A and B in that each element of A corresponds to a unique element of B and *vice versa*: For each $y_0 \in B$ there exists a unique $x_0 \in A$ such that $f(x_0) = y_0$ (Figure 3).

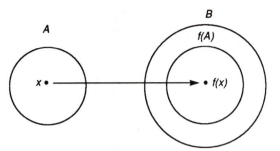

Figure 1

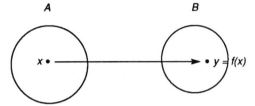

Figure 2

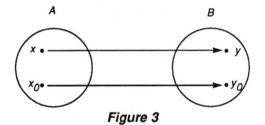

Figure 3

Example 3   Let E be the set of even integers:

$$E = \{0, \pm 2, \pm 4, \pm 6, \ldots\} = \{k \in \mathbf{Z} \mid k = 2n \text{ for some } n \in \mathbf{Z}\}.$$

Define $f : \mathbf{Z} \to E$ as $f(n) = 2n$ for $n \in \mathbf{Z}$. We claim that f is a bijection.

(i) f is injective. Suppose $f(m) = f(n)$ for $m, n \in \mathbf{Z}$. Then $2m = 2n$. By the cancellation property of multiplication, $m = n$. Therefore, f is injective.

(ii) f is surjective. Let $k \in E$. Then $k = 2n$ for some $n \in \mathbf{Z}$ and $f(n) = 2n = k$, hence $k \in \operatorname{Ran}(f)$. Since k is an arbitrary element of E, $\operatorname{Ran}(f) = E$ and f is surjective.

Example 4   Let us take some examples from analytic geometry and calculus.

(i) Define $f : \mathbf{R} \to \mathbf{R}$ by $f(x) = x^2$. Note that $f(2) = f(-2) = 4$, hence f is not one-to-one. Nor is f onto, for $f(x) \neq -1$ for each $x \in \mathbf{R}$. However, if $\mathbf{R}_0 = \{x \in \mathbf{R} \mid x \geq 0\}$, then the function $f : \mathbf{R} \to \mathbf{R}_0$ given by $f(x) = x^2$ is onto. (Given $y \in \mathbf{R}_0$, $y = (\sqrt{y})^2 = f(\sqrt{y})$.) If in addition we restrict the domain of f to the set $\mathbf{R}_0$, then we obtain a function that is both one-to-one and onto: $f : \mathbf{R}_0 \to \mathbf{R}_0$ where $f(x) = x^2$ is a one-to-one correspondence (Figure 4).

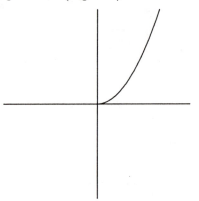

Figure 4

(ii) The sine function $\sin : \mathbf{R} \to \mathbf{R}$, sending x to $\sin(x)$, is neither one-to-one nor onto. However, if $A = [-\pi/2,\, \pi/2]$ and $B = [-1, 1]$, then the sine function is a one-to-one correspondence from A to B (Figure 5).

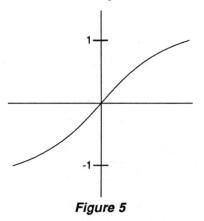

Figure 5

(iii) Let $A = (-\pi/2,\, \pi/2)$. We prove that function $f : A \to \mathbf{R}$ defined by $f(x) = \tan(x)$ is a one-to-one correspondence between A and $\mathbf{R}$, a somewhat surprising conclusion, since the interval A has finite length while the set of real numbers certainly does not. Our argument uses the following facts for whose statements, meanings, and proofs we refer to any text on calculus:

(a) f is continuous on A.
(b) $f'(x) = \sec^2(x) > 0$ for all $x \in A$.
(c) $\lim_{x \to \pi/2+} f(x) = \infty$ and $\lim_{x \to -\pi/2-} f(x) = -\infty$.
(d) The Mean Value Theorem.
(e) The Intermediate Value Theorem.

Proof that f is one-to-one: suppose $x, y \in A$ with $x \neq y$ and $f(x) = f(y)$. Then by the Mean Value Theorem, there is a number c between x and y such that $0 = f(x) - f(y) = f'(c)(x - y)$. Thus, since $x \neq y$, $f'(c) = 0$, a contradiction of (b).

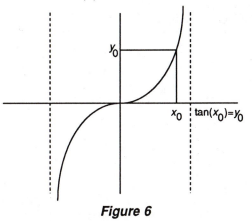

Figure 6

Proof that f is onto: Let $y \in \mathbf{R}$ be an arbitrary positive number. We show that there exists $x \in A$ such that $f(x) = \tan(x) = y$. Since $\lim_{x \to \pi/2+} f(x) = \infty$, there exist $a \in A$ such that $f(a) > y$. Now from the Intermediate Value Theorem, we conclude that there is $x \in A$, $0 < x < a$, such that $f(x) = y$. A similar argument shows that for $y < 0$, there is $x \in A$ such that $f(x) = \tan(x) = y$. The graph of $y = f(x)$ is pictured in Figure 6.

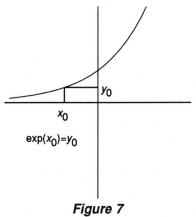

Figure 7

(iv) Let $\exp : \mathbf{R} \to \mathbf{R}^+$ be the exponential function: $\exp(x) = e^x$. In calculus we prove, without explicitly saying so, that exp is a bijection. (See Figure 7 for the graph of this function.)

The figures associated with the above examples suggest the geometric meaning of the bijective property: A function $f : A \to B$ where $A, B \subseteq \mathbf{R}$

is a one-to-one correspondence if and only if for each $y_0 \in B$, the horizontal line passing through $(0, y_0)$ intersects the graph of f in exactly one point (x_0, y_0). Geometric interpretations can also be given individually for the injective and surjective properties.

OPERATIONS ON FUNCTIONS

Definition 5   *Composition of Functions*

   Let $f : A \to B$ and $g : B \to C$ be functions. The *composition of f with g* is the function $g \circ f : A \to C$ defined by $(g \circ f)(x) = g\big(f(x)\big)$ for $x \in A$.

Example 5   For any set A, let $I_A : A \to A$ denote the identity function. Then for any function $f : A \to B$, $f \circ I_A = f$ and $I_B \circ f = f$.

Example 6   Let $\exp : \mathbf{R} \to \mathbf{R}^+$ and $\ln : \mathbf{R}^+ \to \mathbf{R}$ denote the exponential (to base e) function and the natural logarithm function, respectively. Then for $a \in \mathbf{R}$, $\ln(\exp(a)) = \ln(e^a) = a$ and for $b \in \mathbf{R}^+$, $\exp(\ln(b)) = e^{\ln(b)} = b$. Thus, $\ln \circ \exp = I_{\mathbf{R}}$ and $\exp \circ \ln = I_{\mathbf{R}^+}$.

Example 7   Let $f : \mathbf{R} \to \mathbf{R}$ be defined by the equation $f(x) = \sqrt{x^2 + 1}$. Then $f = h \circ g$ where $g : \mathbf{R} \to \mathbf{R}^+$ is given by $g(x) = x^2 + 1$ and $h : \mathbf{R}^+ \to \mathbf{R}$ is defined by $h(x) = \sqrt{x}$.

Composition can be regarded as an operation on functions: To any pair of functions f and g such that $f : A \to B$ and $g : B \to C$, one can associate another function $g \circ f : A \to C$, i.e., each ordered pair of functions (f, g) such that $\mathrm{Ran}(f) \subseteq \mathrm{Dom}(g)$ determine a new function $g \circ f$. When studying the composition of functions, it is natural to ask the following general question: Are properties of f and g inherited by $g \circ f$? In other words, if f and g have a given property, then does $g \circ f$ possess that property? Certainly the most accurate answer one can give is: sometimes. For certain properties the answer might be "yes" for all functions f and g; for other properties, it could be "no" for all f and g; in some cases the answer could be "yes" for some f and g and "no" for others. With regard to injectivity, surjectivity, and bijectivity, the next theorem provides an affirmative answer.

Theorem 1   *Suppose $f : A \to B$ and $g : B \to C$.*
   (i) If f and g are injective, then $g \circ f$ is injective.
   (ii) If f and g are surjective, then $g \circ f$ is surjective.
   (iii) If f and g are bijective, then $g \circ f$ is bijective.
   (iv) (A partial converse to (iii))   If $g \circ f$ is bijective, then f is injective and g is surjective.

Comment   The converse of part (iii) reads: If $g \circ f$ is bijective, then f and g are bijective. However, this statement is false as a simple counterexample shows. (See Exercise 11(b).)

Proof. (i) Suppose $x, y \in A$ with $x \neq y$. We must show that $(g \circ f)(x) \neq (g \circ f)(y)$. Since f is injective and $x, y \in A$, $f(x) \neq f(y)$. But $f(x), f(y) \in B$ and $g : B \to C$ is injective. Therefore, $g(f(x)) \neq g(f(y))$, or equivalently $(g \circ f)(x) \neq (g \circ f)(y)$. We conclude that $g \circ f$ is injective.

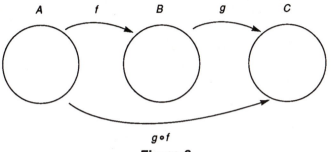

g∘f

Figure 8

(ii) (Informal argument) We must show that each $z \in C$ is "hit" by $g \circ f$, i.e., there exists $x \in A$ such that $(g \circ f)(x) = z$. Now since g is surjective, we can hit z with an element $y \in B$ via the function $g : z = g(y)$ (Figure 8). But since f is surjective, we can hit y from A via $f : y = f(x)$ for some $x \in A$ (Figure 9). It follows that $z = (g \circ f)(x)$.

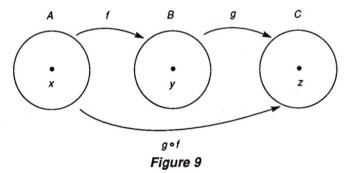

g∘f

Figure 9

(iii) This assertion follows from the definition of bijectivity and parts (i) and (ii).

(iv) We first show that if $g \circ f$ is bijective, then f is injective. We accomplish this goal by showing that for $x, y \in A$, $f(x) = f(y)$ implies that $x = y$. To do this, we must somehow use the assumption that $g \circ f$ is bijective. Since $f(x) = f(y)$ and $f(x) \in B$, $g(f(x)) = g(f(y))$. Therefore, $(g \circ f)(x) = (g \circ f)(y)$. Since $g \circ f$ is injective, it follows that $x = y$. Therefore, f is injective.

We leave the proof that g is surjective as an exercise. ∎

By the way, observe that the proof that f is injective uses only the assumption that $g \circ f$ is injective. Therefore, the following more general result holds: If $f : A \to B$, $g : B \to C$ and $g \circ f$ is injective, then f is injective.

Next we turn to the concept of the inverse of a function.

Definition 6 *Inverse of a Function*

Let $f : A \to B$ be a function. The *inverse of* f is the subset of $B \times A$ defined by

$$f^{-1} = \{(y, x) \in B \times A \mid y = f(x)\}.$$

Note that f^{-1} is a relation from B to A.

Example 8 Let $f : \mathbf{R} \to \mathbf{R}$ be given by $f(x) = x^2$ for $x \in \mathbf{R}$. Then

$$f^{-1} = \{(y, \pm\sqrt{y}) \mid y \in \mathbf{R} \text{ and } y \geq 0\}.$$

Note that $(4, 2) \in f^{-1}$ and $(4, -2) \in f$; hence f^{-1} is not a function from $\{x \in \mathbf{R} \mid x \geq 0\}$ to $\mathbf{R}$. The graphs of f and f^{-1} are depicted in Figure 10.

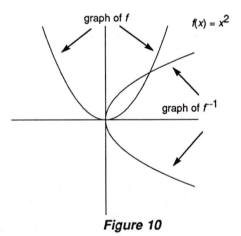

graph of f $f(x) = x^2$

graph of f^{-1}

Figure 10

Example 9 Let $g(x) = x^3$ for $x \in \mathbf{R}$. Thus $g : \mathbf{R} \to \mathbf{R}$ and $g^{-1} = \{(u, \sqrt[3]{u}) \mid u \in \mathbf{R}\}$. Since each real number has a unique real cube root, for any $u \in \mathbf{R}$, there exists exactly one $v \in \mathbf{R}$ such that $(u, v) \in f$. Therefore g^{-1} is a function from $\mathbf{R}$ to $\mathbf{R}$ and the expression $g^{-1}(u) = \sqrt[3]{u}$ gives an unambiguous rule for g^{-1} (Figure 11).

Example 3 (revisited) The function $f : \mathbf{Z} \to E$ defined by $f(n) = 2n$ has an inverse $g : E \to \mathbf{Z}$, which is defined by $g(m) = (1/2)m$, for $(x, y) \in f$ if and only if $y = 2x$ if and only if $x = (1/2)y$ if and only if $(y, x) \in g$. Thus $g = f^{-1}$.

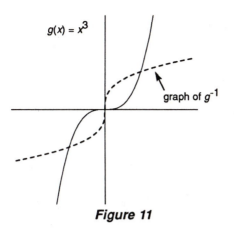

Figure 11

Example 8 shows that for any arbitrary function f, the relation f^{-1} need not be a function. On the other hand, Examples 9 and 3 (revisited) provide examples of functions f for which the relation f^{-1} is a function. A natural question arises: If $f : A \to B$ is a function, under what conditions is f^{-1} also a function? To begin to answer the question, let us glance again at Examples 8 and 9. Why does f fail to be a function? We noticed that since $f(2) = f(-2) = 4$, the pairs $(4, 2)$ and $(4, -2)$ are both in f; hence f^{-1} is not a function. Thus, because f is not injective, f^{-1} is not a function. On the other hand, the function g is injective (g is an increasing function), and it so happens that the relation g^{-1} is a function. These observations and other examples lead to the following theorem.

Theorem 2   *Let A and B be sets and let $f : A \to B$ be a function. Then the relation f^{-1} from B to A is a function if and only if f is injective. In this case, the domain of f^{-1} equals the range of f and the range of f^{-1} is A.*

   Proof. We show that the relation f^{-1} is a function if and only if f is injective. We leave the proof of the second statement as an exercise.
   By definition f^{-1} is a function from B to A if and only if for all $(u, v), (u, w) \in B \times A$, if $(u, v), (u, w) \in f^{-1}$, then $v = w$. This statement is valid if and only if $(v, u), (w, u) \in f$ implies $v = w$. But this statement holds if and only if the function f is injective.   ∎

   Inverse functions play a major role in function theory. Perhaps the most famous example of a function and its inverse are the exponential and logarithmic functions: If $y = \exp(x)$, then $x = \ln(y)$ and conversely if $y = \ln(x)$, then $x = \exp(y)$. Thus, exp and ln are the inverses of each other. Another way of describing these properties of exp and ln is presented in Example 5: $\ln\big(\exp(x)\big) = x$ for all $x \in \mathbf{R}$ and $\exp\big(\ln(x)\big) = x$ for all $x \in \mathbf{R}^+$. In other words, the compositions $\exp \circ \ln$ and $\ln \circ \exp$ are the identities mapping on $\mathbf{R}^+$ and $\mathbf{R}$, respectively. Not surprisingly, the observations noted here for exp and ln hold for any function and its inverse.

Theorem 3 *Let $f : A \to B$ be a bijective function.*
(i) $f : B \to A$ *is also a bijective function.*
(ii) $(f^{-1})^{-1} = f$.
(iii) $f^{-1} \circ f = I_A$ *and* $f \circ f^{-1} = I_B$.

Proof. From Theorem 2 it follows that f^{-1} is a function that is surjective. To show that f^{-1} is injective, suppose $x = f^{-1}(u) = f^{-1}(v)$ where $u, v \in B$. Since $x = f^{-1}(u)$, $(u, x) \in f^{-1}$, which means that $(x, u) \in f$ and $f(x) = u$. Similarly, $f(x) = v$. Therefore, $u = v$ and hence f^{-1} is injective.

To prove (ii) we show that for all $x \in A$, $(f^{-1})^{-1}(x) = f(x)$. (Note that by part (i) $(f^{-1})^{-1}$ is a function from A to B.) Now by the definition of inverse function, $(f^{-1})^{-1}(x) = y$ if and only if $f^{-1}(y) = x$, which holds if and only if $f(x) = y$. Therefore, for all $x \in A$, $(f^{-1})^{-1}(x) = f(x)$.

Finally we prove that $f^{-1} \circ f = I_A$. Letting x be an arbitrary element of A, we show that $(f^{-1} \circ f)(x) = x = I_A(x)$. Let $f(x) = y$. Then

$$(f^{-1} \circ f)(x) = f^{-1}(f(x)) = f^{-1}(y) = x = I_A(x)$$

by the definition of f .   ∎

Let us specialize again to a function $f : \mathbf{R} \to \mathbf{R}$. Then f^{-1} is a relation from $\mathrm{Ran}(f)$ to $\mathbf{R}$. We inquire about the geometric relationship between the graph of f and the graph of f^{-1}. Perhaps the examples thus far presented in this section provide a clue to the answer. In any case, regarding both f and f^{-1} as sets of ordered pairs, we have from the definition of f^{-1} that if $(x, y) \in f$ then $(y, x) \in f$. How are the points (x, y) and (y, x) related geometrically? First suppose $x \neq y$. Then (x, y), (x, x), (y, x), and (y, y) form the vertices of a square (see Figure 12). Therefore, the line segment joining (x, y) to (y, x) is perpendicular to and is bisected by the line ℓ with equation $y = x$. Thus the points (x, y) and (y, x) are symmetric about the line ℓ. If $x = y$, then $(x, y) = (x, x) = (y, x)$ is on ℓ and (x, y) and (y, x) are trivially symmetric about ℓ. The upshot of these remarks is that for any function f, the graph of f^{-1} is obtained by taking the reflection of the graph of f about the line $y = x$.

Consider next a very special case. Suppose $f : \mathbf{R} \to \mathbf{R}$ is a bijective function. Then by Theorem 3 f^{-1} is also a bijective function from $\mathbf{R}$ to $\mathbf{R}$. Suppose $f = f^{-1}$. (In other words suppose $f(x) = f^{-1}(x)$ for all $x \in \mathbf{R}$. Indeed this can happen, an example being provided by the function $f(x) = -x$.) What can be said about the graph of f? By the comments

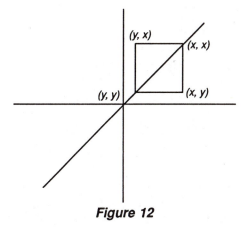

Figure 12

in the previous paragraph, we see that if $f = f^{-1}$, then the graph of f itself is symmetric about the line $y = x$. Conversely, if the graph of a function $f : \mathbf{R} \to \mathbf{R}$ is symmetric about the line $y = x$, then f^{-1} is a function since the graph of f^{-1} (which is the graph of f) is the graph of a function; moreover, $f = f^{-1}$ (see Figure 13).

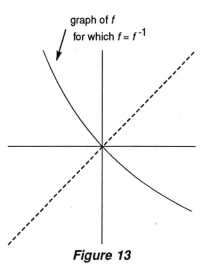

Figure 13

We summarize the results of the last two paragraphs: Suppose $f : \mathbf{R} \to \mathbf{R}$ is a bijection. Then $f^{-1} : \mathbf{R} \to \mathbf{R}$ is also a bijection and $f = f^{-1}$ if and only if the graph of f is symmetric about the line $y = x$.

We complete our discussion of properties of functions with an important technical idea.

Definition 7   *Inverse Image*

    Let $f : A \rightarrow B$ be any function and let $C \subseteq B$. The *inverse image of C under f* or the *preimage of C under f* is the set

$$f^{-1}(C) = \{x \in A \mid f(x) \in C\}.$$

When $C = \{b\}$ where $b \in B$, we write $f^{-1}(b)$ in place of $f^{-1}(C) = f^{-1}(\{b\})$.

    The inverse image of C under f is simply the set of elements of A carried into C by f. For example $f^{-1}(B) = A$. The use of the symbol f^{-1} in $f^{-1}(C)$ might suggest that f is an injective function. However, such need not be the case. For any function $f : A \rightarrow B$ and for any $C \subseteq B$, the subset of A, $f^{-1}(C)$, is defined.

    We describe the behavior of the inverse image relative to the set operations of $\cap$ and $\cup$ in our next result.

Theorem 4   *Let $f : A \rightarrow B$ be a function and let $B_1, B_2 \subseteq B$.*

    (i) $f^{-1}(B_1 \cap B_2) = f^{-1}(B_1) \cap f^{-1}(B_2)$.
    (ii) $f^{-1}(B_1 \cup B_2) = f^{-1}(B_1) \cup f^{-1}(B_2)$.

    Proof. We give a quick proof of (i). We show that $x \in f^{-1}(B_1 \cap B_2)$ if and only if $x \in f^{-1}(B_1) \cap f^{-1}(B_2)$: $x \in f^{-1}(B_1 \cap B_2)$ if and only if $f(x) \in B_1 \cap B_2$ if and only if $f(x) \in B_1$ and $f(x) \in B_2$ if and only if $x \in f^{-1}(B_1)$ and $x \in f^{-1}(B_2)$ if and only if $x \in f^{-1}(B_1) \cap f^{-1}(B_2)$.

    We leave the proof of (ii) as an exercise.   ∎

REPRESENTATION OF FUNCTIONS

    With functions, as with any mathematical object, the problem of representation arises. How can we usefully represent or describe a given function? Experience teaches us rather quickly that the answer depends on the circumstance. On certain occasions a particular kind of representation possesses advantages over others. In situations of even moderate complexity, insight usually occurs when the function (or whatever the mathematical object of concern) is regarded from several points of view. The following discussion presents a variety of viewpoints and serves to summarize a number of the ideas presented in this section.

    1. *Ordered pairs.* To represent a function, we can revert to the definition: a function, f, from A to B is a subset of $A \times B$ with the property that for each $a \in A$, there exists exactly one $b \in B$ such that $(a, b) \in f$. Even though it is given to us in the definition, the ordered pair representation of a function is difficult to work with. First, the dynamic nature of a function, as a transformation of elements of A to elements of B, is

not readily apparent in this representation. Second, the really interesting member of the ordered pair (a, b), namely b, is not strongly emphasized. The primary virtue of the ordered pair approach is in its offering of a clean, set-theoretic definition of the function concept. Most of the time we can operate with descriptions of functions that enable us to manipulate functions easily and accurately.

2. *Equations or Formulas.* When asked to imagine a function f : $\mathbf{R} \to \mathbf{R}$, most of us conjure up an equation that allows us to compute the value of $f(x)$ corresponding to the input x. This computation usually involves a combination of the arithmetic operations of addition, subtraction, multiplication, and division, the algebraic operation of taking roots, and converging infinite series. Examples include

$$f(x) = x^2 + 2x + 3 \qquad\qquad f(x) = \frac{x+1}{x^2+1}$$

$$f(x) = \sum_{n=0}^{\infty} \frac{x^n}{n!}(= e^x) \qquad\qquad f(x) = \sum_{n=0}^{\infty} x^n(= 1/(1-x) \text{ if } |x| < 1).$$

The general classification of functions described here is

(i) Polynomial functions: $a_0 + a_1 x + \ldots + a_n x^n$ where $n \in \mathbf{N}$ and $a_i \in \mathbf{R}$ for $0 \leq i \leq n$.

(ii) Rational functions: $P(x)/Q(x)$ where P and Q are polynomials.

(iii) Algebraic functions: functions that are solutions of algebraic equations whose coefficients are rational functions. For example, if $R(x)$ is a rational function, then $\sqrt[n]{R(x)}$ is an algebraic function since it is a solution of the equation $z^n - R(x) = 0$.

(iv) Analytic functions: $f(x) = \sum_{n=0}^{\infty} a_n x^n$ where $a_i \in \mathbf{R}$.

(v) Fourier series: for a certain class of functions, another type of infinite series representation is important. Let $f : \mathbf{R} \to \mathbf{R}$ be a periodic function with period 1. This means that for all $x \in \mathbf{R}$, $f(x + 1) = f(x)$. Examples of such functions are $\sin(2\pi x)$, $\cos(2\pi x)$, and $\sin(4\pi x)$. Periodic functions arise frequently in the study of natural phenomena. Mathematicians and physicists have long been interested in the problem of expressing an arbitrary periodic function of period 1 in terms of the trigonometric functions $\sin(2\pi x)$, $\cos(2\pi x)$, $\sin(4\pi x)$, etc. Specifically, if f has period 1, there are these constants $a_0, a_1, b_1, a_2, b_2, \ldots$ such that

$$f(x) = a_0 + a_1 \sin(2\pi x) + b_1 \cos(2\pi x)$$

$$= a_2 \sin(4\pi x) + b_2 \cos(4\pi x) + \cdots$$

$$= a_0 + \sum_{n=1}^{\infty} (a_n \sin(2\pi n x) + b_n \cos(2\pi n x)).$$

Such a representation for f is called a Fourier series for f. The theory of Fourier series is concerned with the existence of and properties of Fourier series representation of periodic functions.

Given our repeated exposure to the arithmetic, algebraic, and analytic operations, it is natural for us to think of functions in these terms. Looking at the history of science, one can see the function concept arising in the late Middle Ages and early Renaissance in the work of mathematicians and physicists. From the time of Newton, around 1660, until about 1840, functions were defined to be quantities expressible in one of the forms $1-4$ given above. In the mid-nineteenth century, the function concept was enlarged to include the "correspondence" definition. The ordered pair definition, which had to await the birth of set theory at the turn of the twentieth century, was apparently first formulated in 1939 by N. Bourbaki who is not a real person but actually a pseudonym for a group of French mathematicians.

3. *Tables.* Suppose $f : A \to B$ is a function whose domain is a finite set. Thus $A = \{a_1, \ldots, a_n\}$ where n is a positive integer. Let $b_1 = f(a_1), b_2 = f(a_2), \ldots, b_n = f(a_n)$. Then we can represent f in tabular form:

x	a_1	a_2	$\ldots$	a_n
$f(x)$	b_1	b_2	$\ldots$	b_n

A sequence of real numbers, i.e., a function $s : \mathbf{N} \to \mathbf{R}$, can also be described in a table. Of course, this representation can never be complete. However, a tabular recording of the first few values of the sequence can help in the recognition of a pattern that may govern the values of the sequence.

To illustrate, let us recall Example 3 of Section 7 (p. 64). For $n \geq 1$ let $s_n = 1 + 3 + \cdots + (2n - 1)$ be the sum of the first n odd positive integers. The table for s_n, $n \leq 5$, is

n	1	2	3	4	5
s_n	1	4	9	16	25

On the basis of this table, it appears that $s_n = n^2$ for $n \geq 1$. We can check that indeed for $n = 6, 7, 8, 9, 10$, $s_n = n^2$. Armed with this evidence, one can now proceed to prove by mathematical induction that $s_n = n^2$ for all $n \geq 1$.

4. *Graphs.* If $f : \mathbf{R} \to \mathbf{R}$, then we have the standard pictorial representation of f, the graph of f in the Cartesian plane. The graph of f is the picture in the Cartesian plane of the set $\{(x, f(x)) \mid x \in \mathrm{Dom}(f)\}$. The graph provides a representation of f that is immediately derived from the definition. The close proximity between the definition of and the graph of a function, together with the apparent enhancement that pictures provide to our thinking, make the graph one of the most useful ways of representing a function. Recall that in one-variable calculus, the graph of a function is used to motivate the concepts of derivative (as the slope of a tangent line) and integral (as area under the graph).

5. *Recursive Definitions of Sequences.* Our final example illustrates a way of defining sequences. A sequence $s : \mathbf{N} \to \mathbf{R}$ of real numbers is called *recursive* if there is number n_0 such that for each $n \geq n_0$, s_{n+1} can be expressed in terms of $s_0, s_1, \ldots, s_n$. In other words, a sequence is recursive if any given element in the sequence, beyond a certain point, can be defined in terms of previous members of the sequence.

Example 1 (revisited) (i) The Fibonacci sequence is also recursive for, if $n \geq 2$, then $f_n = f_{n-1} + f_{n-2}$.

(ii) The sequence $\text{fact} : \mathbf{N} \to \mathbf{R}$ defined by $\text{fact}(n) = n!$ is recursive, for whenever $n \geq 1$, $\text{fact}(n + 1) = (n + 1) \cdot \text{fact}(n)$.

Let $s : \mathbf{N} \to \mathbf{R}$ be a recursive sequence and let n_0 be the integer given in the definition of recursivity. Thus $n_0 = 1$ for the factorial sequence and $n_0 = 2$ for the Fibonacci sequence. The numbers $s(0)$, $s(1)$, $\ldots$, $s(n_0-1)$ are called the *initial values* or *initial conditions*. The initial values can be prescribed arbitrarily; thus many distinct sequences exist with the same recurrence relation. For example, if we define $L : \mathbf{N} \to \mathbf{R}$ by $L(0) = 1$, $L(1) = 3$, and $L(n) = L(n - 1) + L(n - 2)$ for $n \geq 2$, we obtain the sequence $1, 3, 4, 7, 11, 18, \ldots$. This close relative of the Fibonacci sequence is called the *Lucas sequence* and is also well-known within mathematics.

Recursive sequences arise frequently in problems that involve the counting of objects. Such problems occur frequently in discrete mathematics including combinatorics and computer science. The idea of a recursive sequence is actually a special case of the more general idea of recursion. Another example is that of a recursive computer program, which is roughly a program that operates by calling "earlier" instances of itself. As you might suspect, recursion is closely related to mathematical induction. A thorough discussion of recursion would take us deep into the realm of set theory, far beyond the scope of this text. Suffice it to say that recursive sequences and recursion in general are ubiquitous themes in mathematics, and as far as sequences are concerned, recursion provides a way of defining and thereby representing sequences.

EXERCISES §12

1. Find all functions $f : A \to B$ when
   (a) $A = \{1, 2\}$ and $B = \{1\}$.
   (b) $A = \{1, 2, 3\}$ and $B = \{1\}$.
   (c) $A = \{a_1, \ldots, a_n\}$ and $B = \{b\}$.
   (d) $A = \{1\}$ and $B = \{1, 2\}$.
   (e) $A = \{1\}$ and $B = \{1, 2, 3\}$.
   (f) $A = \{a\}$ and $B = \{b_1, \ldots, b_n\}$.

2. In each case state whether the given function is injective, surjective, and/or bijective.
   (a) $f : \mathbf{R} \to \mathbf{R}$, $f(x) = 2x$ for $x \in \mathbf{R}$.
   (b) $f : \mathbf{R} \to \mathbf{R}$, $f(x) = 3 - x$ for $x \in \mathbf{R}$.

(c) $f : \mathbf{R} \to \mathbf{R}$, $f(x) = x^2 + 2x + 3$ for $x \in \mathbf{R}$.
(d) $f : [0, \pi) \to [0, 1]$, $f(x) = \sin(x)$ for $x \in [0, \pi)$.
(e) $f : \mathbf{R} \to \mathbf{R}^+$, $f(x) = e^{(x^2)}$ for $x \in \mathbf{R}$.

3. Define $g : \mathbf{Z} \to \mathbf{N}$ as follows:

$$g(x) = \begin{cases} 2x & \text{if } x \geq 0 \\ -2x - 1 & \text{if } x < 0 \end{cases}$$

(a) Evaluate $g(x)$ for $-5 \leq x \leq 5$.
(b) Describe the definition of g in a sentence.
(c) Prove that if $x \in \mathbf{Z}$, then $g(x) \in \mathbf{N}$.
(d) Prove that g is injective.
(e) Prove that g is surjective.
(f) Find g^{-1}.

4. Let $a, b \in \mathbf{R}$ with $a \neq 0$. Define $f : \mathbf{R} \to \mathbf{R}$ by $f(x) = a \cdot x + b$.
(a) Show f is injective.
(b) What is $\mathrm{Ran}(f)$?
(c) Find $f^{-1}(x)$ for $x \in \mathrm{Ran}(f)$.
(d) Sketch the graphs of f and f^{-1}.
(e) As a check of (c), show that if $f(x) = -x$, then $f = f^{-1}$.

5. For $x \in (-\pi/2, \pi/2)$, let $f(x) = \tan(x)$. Sketch the graph of f^{-1}.

6. Show that $\cos \circ \sin : [0, \pi/2] \to \mathbf{R}$ is injective. What is $\mathrm{Ran}(\cos \circ \sin)$?

7. (a) For $n \geq 1$, $n \in \mathbf{N}$, let $S_n = 1 + 2 + \cdots + n$. Give a recursive definition of the sequence S_n.
(b) Give a recursive definition of the sequence t_n where $t_n = 1^2 + 2^2 + \cdots + n^2$.

8. (a) Let $A = \{a, b\}$. How many distinct bijective mappings are there from A to A?
(b) Let $A = \{a, b, c\}$. How many distinct bijective mappings are there from A to A?

9. Let X be any set. Define a relation $\approx$ on $P(X)$ by the rule: $A \approx B$ if there exists a bijective mapping $f : A \to B$. Prove: $\approx$ is an equivalence relation. (Remember that $P(X)$ is the set of all subsets of X, and A and B are subsets of X.)

10. Let $f : A \to B$ with $A_1 \subseteq A$ and $B_1 \subseteq B$.
(a) Prove, or disprove and salvage: $f(f^{-1}(B_1)) = B_1$.
(b) Prove, or disprove and salvage: $f^{-1}(f(A_1)) = A_1$.

11. Let $f : A \to B$ and $g : B \to C$.
(a) Suppose f and g are bijective. Express $(g \circ f)^{-1}$ in terms of f^{-1} and g^{-1}. Prove that your expression is correct.
(b) Give an example to show that $g \circ f$ can be bijective yet neither f nor g need be bijective.

(c) If $g \circ f$ and f are bijective, then is g necessarily bijective?

(d) Prove: If $g \circ f$ and g are bijective, then f is bijective.

12. Prove the second statement in Theorem 2.

13. Prove Theorem 4 (ii).

14. The graph of a function f in the figure below is symmetric about the line $y = x$. What is $f(x)$ for $x \in \mathbf{R}$? What is $f^{-1}(x)$ for $x \in \mathbf{R}$?

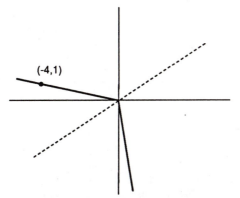

(-4,1)

15. (a) Let $f : \mathbf{R}^+ \rightarrow \mathbf{R}^+$ be defined by $f(x) = 1/x$. Show f is bijective. What is f^{-1}?

(b) Let $g : \mathbf{R}^+ \rightarrow \mathbf{R}^+$ be defined by $g(x) = 1/x^2$. Show g is bijective. What is g^{-1}?

(c) Generalize the statements in (a) and (b) and prove that your generalizations are correct.

16. Let A be a finite set with n elements. Let B_n denote the set of binary sequences of length n. Fill in the blank: The function from $P(A)$ to B_n that assigns each subset of A to its corresponding binary sequence is a _____ function.

17. Let A and B be any sets. Find a one-to-one correspondence $f : A \times B \rightarrow B \times A$.

Let us return to the set $\mathcal{F}$ of fractions of integers mentioned in Example 7 of Section 11: $\mathcal{F} = \{a/b \mid a, b \in \mathbf{Z} \text{ and } b \neq 0\}$. We have all worked with the set $\mathcal{F}$ in both formal mathematics courses and informal applications of mathematics. But no matter what the setting, we do a curious thing. We often use elements of $\mathcal{F}$ to represent quantities—distance, amount, and weight, for example. In these situations, we *regard some fractions that appear to be distinct as being equal.* When regarded as symbols, the fractions $1/2$ and $2/4$ are not identical (the numerators of the two fractions are distinct; so are the denominators); however, the fractions $1/2$ and $2/4$ represent the same quantities. If you have a half dollar and I have two quarters, then you and I have equal amounts of money. Thus when representing quantities, the fractions $1/2$ and $2/4$ are thought of as "equal." In general, the fractions $1/2$ and $a/2a$ where $a \in \mathbf{Z}$, $a \neq 0$, are regarded as equal, and, conversely, if $1/2$ "equals" a/b, then $a \neq 0$ and $b = 2a$.

Coincidentally, this notion of equality of fractions is captured by the relation in Example 7 of Section 11. If $a/b, c/d \in \mathcal{F}$, then $a/b \cong c/d$ if and only if $ad = bc$. For example, it follows that $1/2 \cong a/b$ if and only if $b = 2a$. Therefore, we can say that two fractions a/b and c/d are "equal" if and only if $a/b \cong c/d$. Now to be absolutely proper, we should probably use a word other than "equal" since "equal" is often equated with "identical." Thus we call a/b and c/d *equivalent fractions* if $a/b \cong c/d$, i.e., if $ad = bc$.

The relation $\cong$ is analogous to the identity relation in that $\cong$ possesses the same properties of the identity relation:

1. $\cong$ is a reflexive relation: For each $a/b \in \mathcal{F}$, $a/b \cong a/b$.

2. $\cong$ is a symmetric relation: For each $a/b, c/d \in \mathcal{F}$, if $a/b \cong c/d$, then $c/d \cong a/b$.

3. $\cong$ is a transitive relation: For each $a/b, c/d, e/f \in \mathcal{F}$, if $a/b \cong c/d$ and $c/d \cong e/f$, then $a/b \cong e/f$.

This example leads us to the following general definition.

> **Definition 1** *Equivalence Relation*
>
> A relation $\sim$ on a set A is an *equivalence relation* if $\sim$ is a reflexive, symmetric, and transitive relation on A.

The symbol "$\sim$" is commonly used to denote an equivalence relation, probably because of its resemblance to the symbol "$=$". For $x, y \in A$ we read "$x \sim y$" as "x is *equivalent* to y," or in lighter moments, as "x wiggles y." We write $x \nsim y$ if x is not equivalent to y.

Example 1   As we have noted, the relation $\cong$ on the set $\mathcal{F}$ of fractions of integers is an equivalence relation.

Example 2   Let $\mathcal{F}$ be the set of triangles in the plane. If $T_1, T_2 \in \mathcal{F}$, then we define T_1 Cong T_2 to mean T_1 and T_2 are congruent triangles. (Recall that triangle T_1 is congruent to triangle T_2 if T_2 can be covered precisely by a copy of T_1. See Figure 1.) By an appeal to this definition or to theorems of Euclidean geometry, it is easy to check that Cong is an equivalence relation on $\mathcal{F}$.

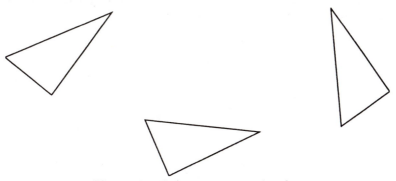

Figure 1   Three congruent triangles.

Hence, equality of fractions and congruence of triangles, two of the most important relations springing from grade school and secondary school mathematics, are in fact equivalence relations. As these examples might suggest, the notion of an equivalence relation is one of the most basic concepts in mathematics. Let us look at equivalence relations on some small sets.

Example 3   (i) Let us determine all equivalence relations on $A = \{a, b\}$. Let $\sim$ be an equivalence relation on A. Since $\sim$ is reflexive, $(a, a), (b, b) \in \sim$; i.e., $\sim \supseteq \{(a, a), (b, b)\} = I_A$. (Here we regard $\sim$ as a subset of $A \times A$.) We have two cases: (1) $\sim = I_A$, and (2) $\sim \supset I_A$. In case (2), either $(a, b) \in \sim$ or $(b, a) \in \sim$. But if $(a, b) \in \sim$, then $(b, a) \in \sim$ by symmetry, and if $(b, a) \in \sim$, then $(a, b) \in \sim$ by symmetry. Therefore, in case (2), $\sim = \{(a, a), (b, b), (a, b), (b, a)\} = A \times A$. To summarize, we have shown that if $\sim$ is an equivalence relation on $A = \{a, b\}$, then either

$\sim \; = I_A$ or $\sim \; = A \times A$.

(ii) As an exercise, find all equivalence relations on $A = \{a, b, c\}$. Here is one example: $\{(a, a), (b, b), (c, c), (a, b), (b, a)\}$.

Example 4   Let R be the relation on the Cartesian plane $\mathbf{R}^2$ defined by

$$(x, y) \, R(u, v) \quad \text{if} \quad x^2 + y^2 = u^2 + v^2.$$

In words, two points in $\mathbf{R}^2$ are related if their respective distances to the origin are equal. A straightforward argument shows that R is an equivalence relation. For instance, to check the transitivity of R, suppose $(x, y) \, R(u, v)$ and $(u, v) \, R(z, w)$; then $x^2 + y^2 = u^2 + v^2$ and $u^2 + v^2 = z^2 + w^2$. By the transitivity of $=$, $x^2 + y^2 = z^2 + w^2$ and $(x, y) \, R(z, w)$. Similarly, the reflexivity (respectively symmetry) of R follows from the reflexivity (respectively symmetry) of $=$. For a given point $(x, y) \neq (0, 0)$, the set of points that are equivalent to (x, y) is a circle centered at $(0, 0)$ of radius $\sqrt{x^2 + y^2}$. See Figure 2.

Thus when looking at $\mathbf{R}^2$ from the point of view of this relation, we pay no attention to the direction from $(0, 0)$ to a point (x, y); what matters is the distance from $(0, 0)$ to (x, y).

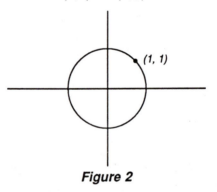

Figure 2

Example 5   As in Example 2, let $\mathcal{F}$ denote the set of triangles in the Euclidean plane. Define a relation Sim on $\mathcal{F}$ by saying T_1 Sim T_2 if triangle T_1 is similar to triangle T_2. (By definition, T_1 is similar to T_2 if a magnification or shrinking of T_1 transforms T_1 into a triangle that is congruent to T_2.) With another appeal to high school geometry, one can show that Sim is an equivalence relation on $\mathcal{F}$.

Examples 2 and 5 drive home the observation that several natural equivalence relations can be defined on a given set. Each equivalence relation isolates a characteristic property relative to which the elements of the set are to be regarded as equivalent or, if you like, relative to which they are to be distinguished. For instance, informally we might say that two triangles are similar if they have the same shape, while two triangles are congruent if they have the same shape and size (see Figure 3).

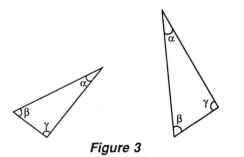

Figure 3

Example 6 Let $\mathcal{L}$ be the set of straight lines in $\mathbf{R}^2$. Define a relation $\|$ on $\mathcal{L}$ by the rule: For $l_1, l_2 \in \mathcal{L}$,

$$l_1 \| l_2 \quad \text{if and only if} \quad l_1 = l_2 \quad \text{or} \quad l_1 \text{ is parallel to } l_2.$$

You can convince yourself that $\|$ is an equivalence relation either formally by referring to theorems of Euclidean geometry or informally be expressing reflexivity, symmetry, and transitivity geometrically. See Figure 4.

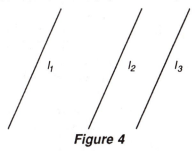

Figure 4

Example 7 Let $A = \mathbf{R}^2 - \{(0,0)\}$. For $(x,y), (u,v) \in A$, define (x,y) Line (u,v) if $(x,y) = (u,v)$ or if the line joining (x,y) to (u,v) passes through the origin. The relation Line can be expressed in the following equivalent fashion:

$$(x,y) \text{ Line } (u,v) \quad \text{if either} \quad x = u = 0 \quad \text{or} \quad y/x = v/u.$$

We leave it as an exercise to check that Line is an equivalence relation.

Example 8 Let E be the relation on $\mathbf{Z}$, the set of integers, defined as follows: $x E y$ if and only if x and y are both even or both odd. We claim that E is an equivalence relation on $\mathbf{Z}$.

(i) E is reflexive: If $x \in \mathbf{Z}$, then $x E x$ since x and x are either both even (if x is even) or both odd (if x is odd).

(ii) E is symmetric: If $x E y$, then x and y are either both even or both odd, and thus y and x are either both even or both odd. Hence $y E x$.

(iii) E is transitive: If $x E y$ and $y E z$, then either x and y are both even in which case (since $y E z$) y and z are both even, and hence x and

z are both even, or x and y are both odd, which implies that y and z are both odd (again since $y\,E\,z$); thus x and z are both odd. Thus, if $x\,E\,y$ and $y\,E\,z$, then $x\,E\,z$.

Therefore, E is an equivalence relation. As far as the relation E is concerned, all that matters about a number is its parity, i.e., whether it is even or odd. (Recall that by the Division Algorithm, if x is an integer, then there exists an integer q such that $x = 2 \cdot q + r$ where $r = 0$ or 1. Thus either $x = 2q$ and x is even, or $x = 2q + 1$ and x is odd.) If x is even, then $x\,E\,0$; if x is odd, then $x\,E\,1$:

$$\{x \mid x\,E\,0\} = \{x \mid x \text{ is even }\} \quad \text{and} \quad \{x \mid x\,E\,1\} = \{x \mid x \text{ is odd}\}.$$

Thus, under the relation E, each element of $\mathbf{Z}$ is equivalent to either 0 or 1.

The next equivalence relation first appeared in the latter half of the eighteenth century and was formally christened in *Disquisitiones Arithmeticae* (Arithmetical Investigations) written by Carl Frederick Gauss (1777–1855) and published in 1801. Gauss is regarded by many as the greatest mathematician of all time because of both the breadth of his work in mathematics, physics, and astronomy and the depth of his insights into every subject that he addressed. For example, in his *Disquisitiones*, Gauss set the principal course for the developments in number theory (which is, to quote TIME magazine, "the abstruse specialty concerned with properties of whole numbers") that were to occur over the following 150 years. Several sections of the *Disquisitiones* are devoted to an analysis of properties of the following concept.

Definition 2 *Congruence Modulo n*

Let n be a positive integer. For $a, b \in \mathbf{Z}$, a *is congruent to* $b \bmod n$ if n divides $a - b$. If a is congruent to $b \bmod n$, then we write either $a \equiv_n b$ or $a \equiv b \pmod{n}$.

Referring back to Example 8, suppose $a\,E\,b$ where $a, b \in \mathbf{Z}$. Then a and b are either both even or both odd, and, in either case, $a - b$ is even, i.e., 2 divides $a - b$. Therefore, if $a\,E\,b$, then $a \equiv_2 b$. Conversely, if $a \equiv b$, then $a\,E\,b$, for otherwise exactly one of a and b is even (and exactly one is odd) and $a - b$ is not divisible by 2. Thus, the equivalence relation E coincides with the relation of congruence modulo 2. The following theorem provides a fairly complete description of the relation $\equiv_n$. Note that its conclusions have already been checked for the relation $\equiv_2$.

Theorem 1 *For each positive integer n, $\equiv_n$ is an equivalence relation. Moreover, each integer is congruent modulo n to exactly one of the elements in the set $\{0, 1, \ldots, n-1\}$.*

Proof. Let n be a positive integer. We show that $\equiv_n$ is an equivalence relation. At first glance, one might be tempted to try an inductive proof of this statement. However it is not clear how in general to relate divisibility by an integer n to divisibility by integers that are less than n. As it happens, a direct proof of the given statement is very straightforward.

(i) Reflexivity: Let $a \in \mathbf{Z}$. Then $a - a = 0 = n \cdot 0$; hence $a \equiv a$.

(ii) Symmetry: Suppose $a \equiv_n b$. We show $b \equiv_n a$. Since $a \equiv_n b$, there exists $k \in \mathbf{Z}$ such that $a - b = nk$. Thus, $b - a = -n \cdot k = n \cdot (-k)$ and $b \equiv_n a$.

(iii) Transitivity: Suppose $a \equiv_n b$ and $b \equiv_n c$. Then $a - b = n \cdot k$ and $b - c = n \cdot m$ for some integers k and m. Hence, $a - c = a - b + b - c = n \cdot k + n \cdot m = n \cdot (k + m)$, and $a \equiv_n c$.

We conclude that $\equiv_n$ is an equivalence relation.

We now prove the second statement. We show that any $a \in \mathbf{Z}$ is congruent modulo n to one and only one element from the set $\{0, 1, \ldots, n - 1\}$.

By the Division Algorithm (p. 66), there exists $q, r \in \mathbf{Z}$ such that $a = n \cdot q + r$ with $0 \leq r \leq n - 1$. Thus, $a - r = n \cdot q$, which implies that $a \equiv_n r$ and $r \in \{0, 1, \ldots, n - 1\}$. Therefore, a is congruent modulo n to at least one element from $\{0, 1, \ldots, n - 1\}$.

Suppose there exist $r, s \in \{0, 1, \ldots, n - 1\}$ such that $a \equiv_n r$ and $a \equiv_n s$. We show $r = s$. Since $\equiv_n$ is an equivalence relation, $r \equiv_n s$ and $r - s = nk$ for some $k \in \mathbf{Z}$. However, $-(n - 1) \leq r - s \leq n - 1$, since $0 \leq r, s \leq n - 1$. Now, if $k = 0$, then $r - s = 0$ or $r = s$. If $k \neq 0$, then by mathematical induction it follows that $|n \cdot k| \geq n$. Hence, either $n \cdot k \leq -n$ or $n \cdot k \geq n$. Thus, if $k \neq 0$, then it is not the case that $-(n - 1) \leq r - s = n \cdot k \leq n - 1$. Therefore, $k = 0$ and $r = s$. (Note the use of an indirect argument to establish the uniqueness portion of the statement.) ∎

A final word on terminology and notation. In all likelihood, Gauss chose the word "congruence" because congruence modulo n is an equivalence relation just as is congruence of triangles. Hence, because of the similarity between the relations $\equiv_n$ and Cong, he borrowed the word "congruence" from geometry to use as a name for $\equiv_n$. As for notation, the symbol $\equiv_n$ has several advantages. First, the subscript reminds us of the modulus, thus preventing possible ambiguity on this score. Second, the symbol $\equiv_n$ suggests the analogy between congruence and equality. Nonetheless, the notation used by Gauss and most mathematicians since is $\equiv$ (mod n). Henceforth, we follow the customs of the mathematical world and write $a \equiv b$ (mod n) to mean a is congruent to b mod n.

EQUIVALENCE CLASSES

We have emphasized that an equivalence relation allows us to broaden the notion of equality from identity to similarity relative to a given property. Two elements need not be identical to be equivalent; they need only to share a specified property. Nevertheless, looking closely at the set $\mathcal{F}$ of fractions, we realize that we usually regard equivalent fractions, such as $1/2$ and $2/4$, as representing the same quantity. Conceptually, we lump together all fractions equivalent to $1/2$ (namely the set $\{a/2a \mid a \in \mathbf{Z}$ and $a \neq 0\}$) and consider them to be a single entity. With the equivalence relation E on $\mathbf{Z}$ defined in Example 8, we identify any two even integers as being equivalent and any two odd integers as being equivalent. Thus the set $\mathbf{Z}$ of integers is split into two subsets (the evens and the odds) and any two elements in the same subset are equivalent. The practice followed in these examples can be generalized and applied to an arbitrary equivalence relation.

Definition 3   *Equivalence Class*

Let $\sim$ be an equivalence relation on a set A. For each $a \in A$, the *equivalence class of* a is the subset, denoted by $[a]_\sim$, consisting of all elements of A that are equivalent to a. In other words,

$$[a]_\sim = \{x \in A \mid x \sim a\}.$$

When there is no ambiguity, we write $[a]$ in place of $[a]_\sim$.

Let us look back at some of our principal examples.

1. $\mathcal{F} = \{a/b \mid a, b \in \mathbf{Z}$ and $b \neq 0\}$ with $a/b \equiv c/d$ if $ad = bc$. Then $[1/2] = \{x \in \mathcal{F} \mid x \equiv 1/2\} = \{a/b \in \mathcal{F} \mid 2a = b\} = \{a/2a \mid a \in \mathbf{Z}$ and $a \neq 0\}$, and $[2/3] = \{a/b \in \mathcal{F} \mid 3a = 2b\} = \{2a/3a \mid a \in \mathbf{Z}$ and $a \neq 0\}$. Notice that $[1/2] = [2/4]$ ($a/b \equiv 2/4$ if and only if $4a = 2b$ if and only if $2a = b$ if and only if $a/b \equiv 1/2$). Therefore, a given equivalence class can be described or represented by several distinct elements.

2. $\mathbf{Z}$ with $a \equiv b \pmod{n}$ if $n \mid a-b$. Consider first the case $n = 2$. Then, writing $[a]_2$ for $[a]_{\equiv_2}$,

$$[0]_2 = \{x \in \mathbf{Z} \mid x \equiv 0 \pmod{2}\} = \{x \in \mathbf{Z} \mid x \text{ is even }\}$$

while

$$[1]_2 = \{x \in \mathbf{Z} \mid x \equiv 1 \pmod{2}\} = \{x \in \mathbf{Z} \mid x \text{ is odd }\};$$

these are the only equivalence classes. Observe that $[0]_2 \cup [1]_2 = \mathbf{Z}$ and $[0]_2 \cap [1]_2 = \emptyset$. These are depicted in Figure 5. Again, each equivalence class can be described in several ways: $[0]_2 = [2]_2 = [4]_2$. The point we emphasize is that the *sets* $[0]_2$, $[2]_2$, and $[4]_2$ are equal; however, their descriptions, as all elements congruent to 0 mod 2 in the first case and all elements congruent to 2 mod 2 in the second case etc., are different.

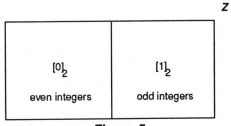

Figure 5

Next consider an arbitrary integer n. We write $[a]_{\equiv_n} = [a]_n$. By Theorem 1, each $a \in \mathbf{Z}$ is congruent to exactly one of the integers $0, 1, 2, \ldots,$ $n - 1$. If $0 \le r \le n - 1$ and $a \equiv r \pmod n$, then $a \in [r]_n$. Also, if $0 \le r,\ s \le n - 1$ and $r \ne s$, then $[r]_n \cap [s]_n = \emptyset$. Thus, there are exactly n equivalence classes of $\mathbf{Z}$ with respect to the congruence modulo n. They are $[0]_n, [1]_n, \ldots, [n - 1]_n$ and $\mathbf{Z} = [0]_n \cup [1]_n \cup \cdots \cup [n - 1]_n$ (Figure 6).

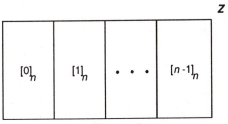

Figure 6

Equivalence classes of $\mathbf{Z}$ with respect to congruence modulo n.

Not surprisingly, the phenomena visible in these examples are quite general.

Theorem 2 *Let $\sim$ be an equivalence relation on a set A and let $x, y \in A$.*
  (i) *If $x \sim y$, then $[x] = [y]$.*
  (ii) *If $x \not\sim y$, then $[x] \cap [y] = \emptyset$.*
  (iii) *The union of the equivalence classes of $\sim$ is A: $A = \bigcup_{x \in A}[x]$.*

  Proof. (i) Suppose $x \sim y$. We show $[x] = [y]$ by proving that $[x] \subseteq [y]$ and $[y] \subseteq [x]$. Let $u \in [x]$. Then $u \sim x$. Since $x \sim y$ and $\sim$ is transitive, $u \sim y$, hence $u \in [y]$. Therefore, $[x] \subseteq [y]$. A "symmetric" argument shows that $[y] \subseteq [x]$. (Just interchange x and y throughout the argument.)

  (ii) We argue by contraposition. We assume $[x] \cap [y] \ne \emptyset$ and show that $x \sim y$. Let $u \in [x] \cap [y]$. Then $x \sim u$ and $y \sim u$. Therefore, by symmetry, $u \sim y$, and by transitivity, $x \sim y$, which contradicts the assumption that $x \not\sim y$.

(iii) The third statement is almost transparent. Each equivalence class is a subset of A. Thus the union of the equivalence classes is a subset of A. On the other hand, each $x \in A$ is in the equivalence class $[x]$, and thus A is contained in the union of the equivalence classes. ∎

Definition 4 *A Set Modulo an Equivalence Relation*

Let $\sim$ be an equivalence relation on A. The set of all equivalence classes is called A *modulo* $\sim$ and is written $A/\sim$.

We emphasize that $A/\sim$ is a set that is formed from the set A and the equivalence relation $\sim$. The elements of $A/\sim$ are certain subsets of A, namely the equivalence classes of A with respect to the equivalence relation $\sim$. Thus $A/\sim \subset P(A)$. The notation $A/\sim$ suggests the process of division, and in a sense, this suggestion mirrors what is actually happening. For the set A is cut up or divided into equivalence classes and each of these sets becomes an element of $A/\sim$.

For the relation $\equiv$ $(\bmod\ n)$ on $\mathbf{Z}$, it is customary to write either $\mathbf{Z}_n$ or $\mathbf{Z}/n\,\mathbf{Z}$ to denote the set of equivalence classes. We will use $\mathbf{Z}/n\,\mathbf{Z}$. By Theorem 1, $\mathbf{Z}/n\,\mathbf{Z}$ has n elements:

$$\mathbf{Z}/n\,\mathbf{Z} = \{[0]_n, [1]_n, \ldots, [n-1]_n\}.$$

PARTITIONS

Looking again at the picture of the equivalence classes of $\mathbf{Z}$ modulo n (Figure 6), we see that the distinct equivalence classes are disjoint and the union of these classes is all of $\mathbf{Z}$. This kind of division of a set has a name.

Definition 5 *Partition*

A *partition* of a set A is a collection P of subsets of A, which are pairwise disjoint and whose union is A.

Note that P is just a collection of subsets of A and is not all of $P(A)$. Also we usually exclude $\emptyset$ from any partition. To say that the sets in P are pairwise disjoint is to say that if $B, C \in P$ with $B \neq C$, then $B \cap C = \emptyset$. The condition that the union of the sets of P is A translates into the equality:

$$A = \bigcup_{B \in P} B$$

Example 9 (i) $P_2 = \{[0]_2, [1]_2\}$ is a partition of $\mathbf{Z}$. The partition P_2 is a set that contains two elements, one of which is the set of even integers and the other of which is the set of odd integers. Note that P_2 is the set of equivalence classes of the equivalence relation $\equiv_2$ on $\mathbf{Z}$.

(ii) $P_n = \{[0]_n, [1]_n, \ldots, [n-1]_n\}$

$$= \{[r]_n \mid r \in \mathbf{Z} \text{ and } 0 \leq r \leq n-1\}$$

is a partition of $\mathbf{Z}$.

These examples suggest a close relationship between equivalence relations and partitions. In fact, Theorem 2 tells us that for any equivalence relation $\sim$ on a set A, the set of *distinct* equivalence classes of A modulo $\sim$ constitutes a partition of A. Thus, *to each equivalence relation on a set* A, *there corresponds a partition of* A. The next theorem assures us that the correspondence runs in the other direction as well.

Theorem 3   *Let P be a partition of a set A. Define a relation $\sim_p$ on A as follows: For $x, y \in A$, $x \sim_p y$ if there is a set B in the partition P such that $x \in B$ and $y \in B$. Then $\sim_p$ is an equivalence relation on A.*

Figure 7 depicts a partition of A and illustrates a pair of related elements and two pairs of unrelated elements.

Proof. We show that $\sim_p$ is reflexive, symmetric, and transitive.

(i) Reflexivity. If $x \in A$, then $x \in B$ for some $B \in P$ (why?), and hence $x \in B$ and $x \in B$; thus $x \sim_p x$, and $\sim_p$ is reflexive.

(ii) Symmetry. If, for $x, y \in A$, $x \sim_p y$, then there is $B \in P$ such that $x \in B$ and $y \in B$; hence $y \in B$ and $x \in B$ and $y \sim_p x$. Thus $\sim_p$ is symmetric.

(iii) Transitivity. Suppose for $x, y, z \in A$, $x \sim_p y$ and $y \sim_p z$. Then there exist $B, C \in P$ such that $x \in B$, $y \in B$, $y \in C$, and $z \in C$. Therefore, $y \in B \cap C$, which means that, since P is a partition of A, $B = C$. Thus, there is a set $B \in P$ such that $x \in B$ and $z \in B$. Therefore, $x \sim_p z$ and $\sim_p$ is transitive.   ∎

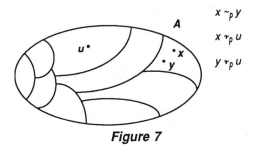

$$x \sim_p y$$
$$x \not\sim_p u$$
$$y \not\sim_p u$$

Figure 7

Example 10   Consider the set $\mathbf{R}$ of real numbers. A partition of $\mathbf{R}$ can be obtained by defining for each $i \in \mathbf{Z}$, $A_i = \{x \in \mathbf{R} \mid i \leq x < i+1\}$. Thus $A_0 = \{x \in \mathbf{R} \mid 0 \leq x < 1\}$ and $A_{-5} = \{x \mid -5 \leq x < -4\}$. Then $P = \{A_i \mid i \in \mathbf{Z}\}$ is a partition of $\mathbf{R}$. What is the corresponding equivalence relation? If $x, y \in \mathbf{R}$, then $x \sim_p y$ if and only if there exist

$A_i \in P$ such that $x \in A_i$ and $y \in A_i$ if and only if there exists $i \in \mathbf{Z}$ such that $i \leq x, y < i+1$. This relation can be expressed using a commonly used function: For $a \in \mathbf{R}$, define $\lfloor a \rfloor$ to be the largest integer $\leq a$. Then $\lfloor 1/2 \rfloor = 0$, $\lfloor -1/2 \rfloor = -1$, and $\lfloor n \rfloor = n$ if $n \in \mathbf{Z}$. The function $\lfloor a \rfloor$ is called either the *greatest integer function* evaluated at a or the *floor function* evaluated at a. Using this concept, we can state that $x \sim_p y$ if and only if $\lfloor x \rfloor = \lfloor y \rfloor$. This example illustrates how one can begin with a partition, use it to define an equivalence relation, and then find another way of describing the equivalence relation.

A summary of Theorems 2 and 3 might be in order. The thrust of Theorem 2 is that each equivalence relation $\sim$ on a set A determines a partition, let's call it $P(\sim)$, of A: $P(\sim)$ is just the set of distinct equivalence classes of A modulo $\sim$. On the other hand, by Theorem 3, each partition P of A determines an equivalence relation $\sim_p$ on A.

Example 11 (The Rational Numbers) We close this section by using the ideas of equivalence relation and equivalence classes to define the set of rational numbers. In effect we are presenting a formal version of Example 1 of this section. We take the set $\mathbf{Z}$ of integers as given: $\mathbf{Z} = \{0, \pm 1, \pm 2, \pm 3, \ldots\}$. Let A be the set $A = \mathbf{Z} \times (\mathbf{Z} - \{0\})$. Thus

$$A = \{(a, b) \mid a, b \in \mathbf{Z} \text{ and } b \neq 0\}.$$

On A, define the relation $\cong$ by $(a, b) \cong (c, d)$ if $ad = bc$. Then $\cong$ is an equivalence relation on A.

The set A is a formal way of describing the set $\mathcal{F}$ of fractions of integers and the equivalence relation $\cong$ exactly expresses the notion that two fractions are equivalent. Now we define the set of rational numbers, $\mathbf{Q}$, to be the set $\mathbf{Q} = A/\cong$. In other words, $\mathbf{Q}$ is the set of equivalence classes of A with respect to the equivalence relation $\cong$. Notice that to define $\mathbf{Q}$ all we needed was the set of integers, $\mathbf{Z}$, and the concept of multiplication on $\mathbf{Z}$, and the notion of equivalence relation.

In Exercise 8 below, we show how one can define the set of integers, $\mathbf{Z}$, given the set of natural numbers, $\mathbf{N} = \{0, 1, 2, \ldots\}$.

EXERCISES §13

1. (a) How many equivalence relations are there on $\{a, b, c\}$?

   (b) Give four examples of equivalence relations on $\{a, b, c, d\}$.

2. Let $\mathcal{L}$ be the set of lines in the Euclidean plane. For $l_1, l_2 \in \mathcal{L}$, define $l_1 \perp l_2$ to mean l_1 and l_2 are perpendicular. Is $\perp$ an equivalence relation?

3. (a) Let A be any set. Which equivalence relation(s) on A deserve the label "trivial"? Why?

   (b) Characterize those sets A that have only one equivalence relation.

4. Describe two equivalence relations on $\mathcal{T}$, the set of triangles in the Euclidean plane, which are distinct from Cong and Sim.

5. Which of the following relations R are equivalence relations on $\mathbf{Z} - \{0\}$?

   (a) $a\,R\,b$ if $2 \mid (a+b)$
   (b) $a\,R\,b$ if $3 \mid (a+b)$
   (c) $a\,R\,b$ if $a \mid b$ and $b \mid a$
   (d) $a\,R\,b$ if $a \mid b$ or $b \mid a$
   (e) $a\,R\,b$ if $a \mid b$

6. Let $\mathbf{P}$ be the set of all propositions in English. For $p, q \in \mathbf{P}$ define $p \sim q$ if $p \Leftrightarrow q$ is true. Show: $\sim$ is an equivalence relation. In other words, logical equivalence is an equivalence relation on the set of English propositions.

7. Prove, or disprove and salvage: Let $[a, b] = \{x \in \mathbf{R} \mid a \le x \le b\}$. Let $C[a, b]$ denote the set of continuous functions on $[a, b]$. If $f, g \in C[a, b]$, then define $f \sim g$ if $\int_a^b f(x)\, dx = \int_a^b g(x)\, dx$. Then $\sim$ is an equivalence relation on $C[a, b]$.

8. Let $\mathbf{N} = \{0, 1, 2, 3, \ldots\}$ be the set of natural numbers. We will use $\mathbf{N}$ to define the set $\mathbf{Z}$ of integers. In other words, given 0 and the positive integers, we want to define the negative integers. Intuitively, a negative integer arises when one subtracts from a given natural number a larger natural number: $2 - 5 = -3$ or $1 - 4 = -3$. Thus each negative integer is determined by a pair of natural numbers, although different pairs can determine the same integer. Let $A = \mathbf{N} \times \mathbf{N}$. For $x, y \in A$ with $x = (a, b)$ and $y = (c, d)$, define $x \sim y$ if $a + d = b + c$. Thus, for example, $(2, 5) \sim (1, 4)$.
   (a) Prove: $\sim$ is an equivalence relation on A.
   (b) The set $\mathbf{Z}$ of integers is then defined to be the set $\mathbf{Z} = A/\!\sim$.

9. (a) Devise an equivalence relation on $\mathbf{R}$ with exactly two equivalence classes.
   (b) Devise an equivalence relation on $\mathbf{R}$ with exactly three equivalence classes.

10. Define the relation $\cong_1$ on $\mathbf{R}$ by the rule: If $x, y \in \mathbf{R}$, then $x \cong_1 y$ if $x - y$ is an integer.
    (a) Show that $\cong_1$ is an equivalence relation on $\mathbf{R}$.
    (b) What is $[1/2]_{\cong_1}$?

11. Define a relation R on $\mathbf{R}^2$ as follows: For $v, w \in \mathbf{R}^2$ with $v = (x, y)$ and $w = (a, b)$, $v\,R\,w$ if $|x| + |y| = |a| + |b|$.
    (a) Prove that R is an equivalence relation on $\mathbf{R}^2$.
    (b) Describe the equivalence classes geometrically.

12. Let $A = \mathbf{Z}^2 = \mathbf{Z} \times \mathbf{Z}$. For $(a, b), (c, d) \in A$, we write $(a, b) \equiv_2 (c, d)$ if $a - c$ is even and $b - d$ is even. Show $\equiv_2$ is an equivalence relation on A. Describe the partition of A corresponding to $\equiv_2$.

13. Let $A = \mathbf{Z}^+ \times \mathbf{Z}^+$. ($\mathbf{Z}^+$ is the set of positive integers.) Define a relation $\sim$ on A by $(a, b) \sim (c, d)$ if $a^b = c^d$.

(a) Show $\sim$ is an equivalence relation on A.

(b) Find the equivalence classes $[(16,1)]_\sim$ and $[(3,4)]_\sim$.

(c) Find an equivalence class with exactly four elements.

(c) Find an equivalence class with infinitely many elements.

14. Let $A = \mathbf{N} \times \mathbf{N}$. Define a relation on A by $(a,b) \sim (c,d)$ if $ab = cd$.

(a) Show $\sim$ is an equivalence relation on A.

(b) Find an equivalence class with exactly one element.

(c) Find an equivalence class with exactly three elements.

(d) Find an equivalence class with infinitely many elements.

Unit 3
SOLVING PROBLEMS

The heart of mathematics is its problems.

P. R. Halmos

Section 14
HOW TO SOLVE IT

If Paul Halmos, a well-known twentieth-century mathematician, is indeed correct, then the ability to solve mathematical problems, especially those that differ appreciably from any previously encountered exercise, is the foremost skill that a mathematician must possess. The typical person on the street probably believes that this ability is granted at birth to a chosen few and denied to the rest of the population. According to the popular folklore, either you are a mathematical genius or you can't do math at all.

On closer inspection, however, the simplicity of this viewpoint becomes apparent. Mathematical aptitudes and abilities vary over a seemingly continuous range. Some people prefer algebra to geometry in high school; others enjoy geometry far more than algebra. Some find high school mathematics easy and college mathematics difficult. Some are attracted more by "pure" mathematics than by "applied" mathematics. Others choose computational mathematics over "theorem-proving" mathematics. Moreover, as people study and practice mathematics, they generally do become more adept at learning new mathematical topics and more skillful at solving problems. In addition, many of us have encountered an inspiring mathematics teacher who aroused our interest and increased our performance in mathematics beyond what we thought possible. While no one can deny the existence of innate mathematical differences among individuals, it is clear that environmental influences and personal characteristics are also important in determining the mathematical performance of each individual.

Our question is, then: What can we as individuals do to improve our own mathematical performance? Certainly we can learn as much mathematics as possible. For example, the more we know, the more likely we are to be able to answer a particular question. But beyond that, how can we become better at solving problems, proving theorems, writing computer programs, and building mathematical models of phenomena? In short, how can we become better at doing mathematics?

The usual answer to this question is: practice. If you sit down and try to solve many mathematical problems or to prove many theorems, eventually you will become good at problem solving or theorem proving. The experience that you gain from failure, partial success, and success teaches you, perhaps slowly and painfully, how to do mathematics. But is "practice and experience" the only answer to these questions?

Our response to this question is a qualified "no." Practice and the experience gained from practice are necessary for the improvement of one's mathematical performance. But they are not sufficient to guarantee acceptable results. In fact, we assert that proficiency in mathematics requires

1. knowledge of subject matter,
2. awareness of strategies and tactics for solving problems and learning new material,
3. ability to monitor and to control one's activities when doing mathematics.

Most mathematical textbooks have justifiably been concerned only with item 1. The purpose of this unit and the next, however, is to focus on items 2 and 3. Our goal is to introduce and discuss briefly some of the considerations involved in 2 and 3.

What we present are some fundamental rules-of-thumb that might be of assistance in the problem solving process. No one of the principles is guaranteed to work on any given problem. Nevertheless, they can serve to help a person discover a solution to a problem or a proof of a theorem. Guidelines of this sort—those that aim to help in the discovery of a solution but that are not certain to lead to a solution—are called *heuristic* principles. We now turn to a discussion of the heuristics, i.e., the rules of discovery, of problem solving.

Our exposition is based on the work of George Polya and Alan Schoenfeld. Both Polya and Schoenfeld are professional mathematicians who have devoted considerable effort to the study of the activity of mathematical problem solving by humans. Polya's principal method was introspection: He carefully observed his own actions as he was solving mathematical problems, noting particularly the steps that enabled him to solve a difficult problem. In a well-known book, *How To Solve It*, Polya organized his findings into a scheme that divides the problem-solving processing into four parts:

1. Understand the Problem
2. Devise a Plan
3. Carry Out the Plan
4. Look Back

In contrast to Polya, Schoenfeld has carried out a number of controlled experiments with college students and professional mathematicians. His work has led him to modify Polya's scheme somewhat. In Schoenfeld's scheme there are three major steps:

1. Analysis
2. Exploration
3. Verification

We present a hybrid version of the Polya and Schoenfeld schemes.

I. Understand the Problem (Analysis)
   1. Identify the unknown.
   2. Isolate the hypotheses and the data.
   3. Develop a representation of the problem.
      a. Draw a figure.
      b. Introduce suitable notation.
   4. Examine special cases.
      a. Select special values to acquire a "feel" for the problem.
      b. Consider extreme cases.
      c. Evaluate integer parameters at $n = 1, 2, 3, \ldots$ and look for a pattern
   5. Simplify the problem.
      a. Exploit symmetry.
      b. Choose appropriate units.

II. Devise a Plan (Exploration)
   1. Consider essentially equivalent problems.
      a. Replace conditions by equivalent conditions.
      b. Recombine the elements of the problem in various ways.
      c. Introduce auxiliary elements.
      d. Reformulate the problem.
         i. Change perspective or notation.
         ii. Argue by contradiction or contrapositive.
         iii. Take the problem as solved; i.e., assume you have a solution and determine its properties.
   2. Consider slightly modified problems.
      a. Aim for subgoals.
      b. Relax a condition, then restore it.
      c. Consider case analysis.

3. Consider broadly modified problems.
    a. Construct an analogous problem with fewer variables.
    b. Generalize the problem.
    c. Hold all but one variable fixed to determine that variable's impact.
    d. Try to exploit any problem that has a similar form, hypothesis, or conclusion.

III. Carry out the Plan (Verification)
  1. Check each step.
  2. Prove that each step is correct.

IV. Look Back (Verification)
  1. Apply these specific tests to your solution:
      a. Does it use all the pertinent data?
      b. Does it conform to reasonable estimates or predictions?
      c. Does it withstand tests of symmetry, dimension analysis, or scaling?
  2. Apply these general tests:
      a. Can the result be obtained differently?
      b. Can the result be verified in special cases?
      c. Can the result be reduced to known results?
      d. Can the result be used to derive other known results?

The remainder of this unit will be devoted to comments on certain aspects of this outline. More details can be found in Polya's books *How to Solve It* [6], *Mathematical Discovery* [7], and *Mathematics and Plausible Reasoning* [8] and in Schoenfeld's *Mathematical Problem Solving* [9].

Section 15

UNDERSTANDING THE PROBLEM

FOCUSING ON THE UNKNOWN

When presented with a problem, the problem solver's first task is to understand the problem. Understanding a problem (or concept) involves several activities and stages; however, Polya's clear and forceful advice is to begin with the unknown. In Polya's chart (*How to Solve It*, p. xvi–xvii) the first line begins with the question: What is the unknown? Focusing on the unknown provides coherence to the problem and directs the efforts of the problem solver towards a specific aim. In other words, if you know what you are looking for, then you are free to delve deeper into other aspects of the problem in order to obtain a more thorough understanding of it.

The unknown itself can take various forms. For example, it can be a specific mathematical object—a real number, a point in the plane, the point of intersection of two lines, a certain function, or a circle. In the latter case, for instance, to find the unknown circle, one must find its center (a point in the plane) and its radius (a positive real number). In other instances the unknown is not so much a specific object as a task to perform or a goal to achieve: Prove that $\sqrt{2}$ is irrational, or find an efficient algorithm that will yield the solution to a given system of equations.

Whether looking for a specific object or a general task, by focusing on the unknown, the problem solver is forced to begin with the goal and to understand the meaning of all the terms in the unknown. What must be done to find a circle? What is an irrational number? As obvious as it might seem, the advice to focus on the unknown and to know the meaning of all the terms involved in the unknown (and the known, for that matter) is often overlooked. Let us look at an example.

Example 1 How many squares are there on an ordinary checkerboard?

The standard checkerboard is a square, each side of which is divided into eight segments all of the same length. The grid formed by these segments consists of 64 squares. The question asks for the number of squares on such a checkerboard.

One might be tempted to answer 64, but this is the number of "1×1" unit squares. Many other squares can be found in the checkerboard, squares whose sides consist of two unit segments or three unit segments, etc., up to eight unit segments. For instance, there is only one 8 × 8 square on the standard checkerboard. To sum up, our unknown is the total number of squares on a checkerboard. To find this number we must count the number of 1 × 1 squares, 2 × 2 squares, and so on up to the one 8 × 8 square.

Example 2   Find the center and radius of the circle that has equation

$$x^2 + 8x + y^2 - 4y - 6 = 0. \tag{$*$}$$

Recall that a circle consists of all points that are a fixed distance (the radius) from a given point (the center). Thus the unknowns are the radius, r, and the center (a, b).

Now how can we find these unknowns? To answer this question, remember that by the distance formula, the circle consists of all points (x, y) such that $\sqrt{(x-a)^2 + (y-b)^2} = r$ or equivalently

$$(x - a)^2 + (y - b)^2 = r^2. \tag{$**$}$$

Therefore, to find (a, b) and r, it is reasonable to try to transform equality $(*)$ into an equality that has the same form as $(**)$. If we succeed, then we will have found the center (a, b) and the radius r.

Notice that in Example 2 we first identified the unknowns and then took the further step of pinpointing the meaning of the unknowns. In doing so, we have drawn the problem into sharper focus: using equality $(*)$, we must find real numbers a, b, and r that satisfy equality $(**)$. Thus by asking for the meaning of the unknown, we develop a clearer understanding of the problem. So, whenever you encounter a problem, always ask yourself the following questions:

> What is the unknown?
> What do the terms in the unknown mean?

As an exercise, take any mathematics book that you used in a previous course. Open the book to any section that you studied in that course and read over the exercises given at the end of the section. In each case identify the unknown and the meaning of the unknown. Repeat this procedure for several sections of the text and for the exercises in Unit II of this book. Concentrate only on the identification and the meaning of the unknown. Ignore the techniques used in solving the problems. Does concentrating only on the unknown make the problems seem clearer and less forbidding now than when you first encountered them?

George Polya has much more to say on the issue of focusing on the unknown. See the section "Look at the Unknown" in the "Dictionary" portion of *How to Solve It* [6]. We need not repeat Polya's remarks here. However, the moral of the story does bear repeating: In any problem, identify the unknown and understand its meaning.

ISOLATE THE HYPOTHESIS

Once you have a clear understanding of the unknown, focus your attention on the material that is given in the problem, the hypothesis and data. You should approach the hypothesis in the same way that you approach the unknown—with questions:

What is given in the problem?
What is the hypothesis?
What are the data?
What do the terms in the hypothesis mean?

Let us look at a couple of examples.

Example 3   Show that a nonzero polynomial with real coefficients has a finite number of real roots.

Restating the assertion as an implication, we must show that if a polynomial has real coefficients, then that polynomial has only a finite number of real roots. In this problem we are given a polynomial, call it $P(x)$, with real coefficients. This means that $P(x)$ has the form:

$$P(x) = a_n x^n + a_{n-1} x^{n-1} + \cdots + a_1 x + a_0$$

where $a_n, a_{n-1}, \ldots, a_1, a_0$ are real numbers, not all of which are 0. Thus $P(x)$ can be thought of as a function from the set $\mathbf{R}$ to itself that is given by a specific formula such as the one given above. Just for the record, we note that the conclusion of the implication, in this case the unknown in the problem, asserts that $P(x)$ has only a finite number of real roots, i.e., there are only a finite number of real numbers $y_1, \ldots, y_m$ such that $P(y_i) = 0$ for $1 \le i \le m$.

Example 4   A male goat is attached by a rope 8 yards long to a corner of a rectangular shed whose dimensions are 5 yards by 7 yards. What is the area of the patch of grass that the goat can graze?

In this problem, the given material consists of the shape and dimensions of the shed and the length of the goat's rope. We are also given the sex of the goat, but this information is apparently irrelevant to the question.

PROBLEM REPRESENTATION

Suppose that in a given problem you have isolated the unknown and the hypotheses or conditions. Suppose also that the meanings of all the terms involved in the problem are clear. What next? At this point the formulation of an adequate representation of the problem and its components will probably be most useful. By a representation of the problem, we mean a method of describing the problem that

(i) is internally coherent or consistent,

(ii) corresponds closely to the problem,

(iii) is closely connected with the knowledge of the problem solver about the items of the problem and related concepts.

The representation might take any one of several forms. It might involve a picture, a graph, an algebraic equation or a system of algebraic equations, a differential equation or a system of differential equations, a vector diagram, or a mathematical system such as a vector space, a group, or a Boolean algebra. In summary, a representation of a problem amounts to a view of the problem that is sensible, relevant, and can be manipulated in order to achieve a solution. Let us consider some examples of representations.

Perhaps the most familiar kind of representation is the pictorial or geometric representation of algebraic equations involving real numbers. We reconsider some previous examples to illustrate how a picture can be helpful.

Example 5 If x is a real number and $x^2 - 6x + 10 = x$, then $x = 2$ or $x = 5$. (See Section 4, Example 2.) To represent this problem pictorially, recall that the equation $y = x^2 - 6x + 10$ has as its graph in the plane the parabola, P, with vertex at the point $(3,1)$ and focus at $(3,2)$. (The line $x = 3$ is the axis of symmetry of the parabola.) The graph of $y = x$ is the straight line, L, with slope 1 passing through the origin. The values of x for which $x^2 - 6x + 10 = x$ are precisely the first coordinates of the points at which the curves P and L intersect. The picture (Figure 1) makes the result appear extremely plausible. In fact, on the basis of the picture, we might

(i) guess that $x = 2$ and $x = 5$ are solutions to the original question,

(ii) check that $x = 2$ and $x = 5$ actually satisfy the equation,

(iii) show that no other solutions exist.

On the other hand, even if the picture is not used (in this or any similar question) as a means of discovering solutions to the equation, it can be used as a check on a solution obtained by another method.

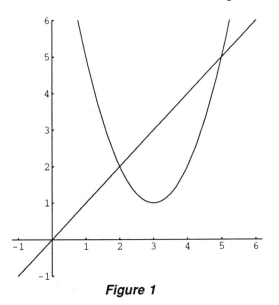

Figure 1

Example 6   Show that for each positive integer n, $1 + 2 + \cdots + n = n(n+1)/2$. (See Section 7, Example 1.)

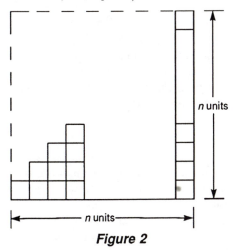

Figure 2

How can an equality of this sort be pictured? In this case we view the numbers on the left and right of the equality as different representations of a given geometric quantity. Of the common geometric quantities—length, area, volume—we consider area.

Can we represent $1 + 2 + \cdots + n$ as the area of a geometric figure? The sum suggests that we find figures of area $1, 2, \ldots, n$, respectively; then the area of the totality of these figures is $1 + 2 + \cdots + n$. A simple figure of area 1 is a square, so let us consider squares arranged in the step-like fashion shown in Figure 2. The area of this figure is indeed $1 + 2 + \cdots + n$. Note that the figure sits inside an $n \times n$ square and apparently covers slightly more than half the square. Therefore,

$$1 + 2 + \cdots + n > n^2/2.$$

Since $n(n+1)/2$ is also slightly greater than $n^2/2$, the equality $1 + 2 + \cdots + n = n(n+1)/2 = n^2/2 + n/2$ is at least plausible.

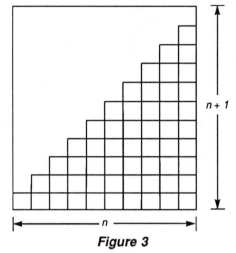

Figure 3

A minor modification of this geometric representation makes the equality even more plausible. The rectangle in Figure 3 has area $n(n+1)$ and is divided into two "step-like" regions of equal area, namely the gridded region and the nongridded region; the area of each figure is therefore $n(n+1)/2$. On the other hand, the area of the lower region is $1 + \cdots + n$. Hence the equality $1 + \cdots + n = n(n+1)/2$.

Drawing a figure provides one method for representing a mathematical problem or situation. However, there are other kinds of representation. For example, introducing symbols to stand for unknown quantities constitutes a form of representation. In max-min or related rate problems in calculus, one obtains a functional relationship among two or more relevant variables which constitutes in effect a representation of the problem.

In some cases a representation entails the introduction of a new idea. For instance, the system of linear equations

$$ax + by = e$$

$$cx + dy = f$$

(a, b, c, d, e, and f are given real numbers; x and y are unknown real numbers that, if they exist, satisfy both equations) can be *represented* in matrix form $AX = B$ where

$$A = \begin{bmatrix} a & b \\ c & d \end{bmatrix}, \qquad X = \begin{bmatrix} x \\ y \end{bmatrix}, \qquad \text{and } B = \begin{bmatrix} e \\ f \end{bmatrix}.$$

To describe this representation we need the notions of matrix and matrix multiplication. The advantage of the matrix representation is that it suggests an analogy with real numbers: The matrix equation $AX = B$ is like the equation $ax = b$ where a and b are given real numbers and x is an unknown real number. From this analogy it is natural to consider the idea of an invertible matrix and to solve the matrix equation in case A is invertible. The solution to the matrix equation can then be interpreted to yield a solution to the original linear system.

EXAMINE SPECIAL CASES

Often one can increase one's understanding of a problem or question by examining special cases. What does this mean? A *special case* of a problem is a particular instance of the problem obtained by considering specific *simple* values for some or all of the variables in the problem. For example, if one of the variables is a positive integer n, then consider the cases $n = 1, 2, 3$; if one of the variables is a real number ranging over the interval $-1 \leq x \leq 1$, then consider the cases $x = \pm 1$, $x = 0$; if one of the variables is an arbitrary triangle T, then investigate the cases in which T is an equilaterial, isosceles, or right triangle. By considering special cases, we can at least acquire some experience with the problem and develop a feel for various aspects of it. Also, by looking at special cases, we can perhaps obtain an idea of how to solve the problem. Our hope is that the consideration of special cases will give us understanding and insight. We will illustrate the strategy of considering special cases with several examples.

A word of forewarning is appropriate: The advice to examine special cases is very general and thus can be difficult to carry out effectively. In a given problem there might be several distinct ways of forming special cases; some of these formations might be helpful in solving the problem, while others might not be.

Example 7   For each positive integer n, evaluate

$$a_n = \sum_{k=1}^{n} k/(k+1)! = 1/2! + 2/3! + \cdots + n/(n+1)!$$

(Recall that for $k \in \mathbf{N}$, $k! = k(k-1) \cdot \cdots \cdot 2 \cdot 1$. The first few values of $k!$ are $1! = 1$, $2! = 2$, $3! = 6$, $4! = 24$, $5! = 120$, $6! = 720$.) Consider the special cases $n = 1, 2, 3, 4, 5$. We have $a_1 = 1/2$, $a_2 = 5/6$, $a_3 = 23/24$, $a_4 = 119/120$, and $a_5 = 719/720$. Thus $a_1 = 1/2!$, $a_2 = 5/3! = 3! - 1/3!$, $a_4 = 119/5! = (5! - 1)/5!$, and $a_5 = 719/6! = (6! - 1)/6!$. Thus it appears that $a_n = ((n+1)! - 1)/(n+1)! = 1 - 1/(n+1)!$.

Example 3 (revisited)   Show that a nonzero polynomial with real coefficients has a finite number of real roots.

Let $P(x)$ be a nonzero polynomial with real coefficients. Then there are real numbers $a_0, \ldots, a_n$, not all 0, such that $P(x) = a_0 + a_1 x + \cdots + a_n x^n$. Let us assume that $a_n \neq 0$. Then the integer n is called the *degree* of $P(x)$. Consider the case $n = 0$. Then $P(x) = a_0$ is a nonzero constant polynomial. Thus $P(r) \neq 0$ for all $r \in \mathbf{R}$ and $P(x)$ has no real roots. Next let $n = 1$. Then $P(x) = a_0 + a_1 x$ where $a_1 \neq 0$, and $P(r) = 0$ if and only if $r = -a_0/a_1$. Hence, $P(x)$ has one real root. Now let $n = 2$. Then $P(x) = a_0 + a_1 x + a_2 x^2$. There are at least a couple of ways of analyzing the roots of $P(x)$; we will take a geometric approach. By analytic geometry, the graph of $P(x)$ is known to be a parabola whose axis is perpendicular to the x-axis. Thus the graph of $P(x)$ has one of the forms shown in Figure 4.

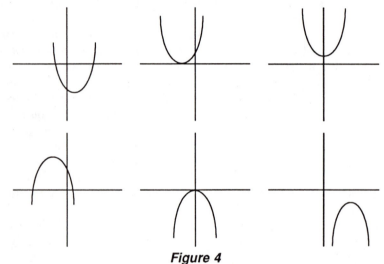

Figure 4

Recall that for a real number r, $P(r) = 0$ if and only if the graph of $P(x)$ crosses the x-axis at the point $(r, 0)$. Hence, geometrically, we see that a polynomial of degree 2 has either zero, one, or two real roots. In fact, the three special cases $n = 0, 1, 2$ suggest a sharpening of the original statement: Prove that a nonzero polynomial with real coefficient having degree n has at most n real roots. Thus the examination of special cases leads us to a more precise formulation of the original problem.

Example 8   Show that for any positive integer n, the product of n consecutive positive integers is divisible by $n!$.

There are at least two ways of considering special cases of this problem. Perhaps the most obvious way is to take $n = 1, 2, 3$. For example, for $n = 2$, one must show that the product of any two consecutive positive integers is divisible by $2! = 2$. This statement is true since one of the two consecutive integers is even and hence the product of the two integers is also even, i.e., is divisible by 2. A second way of considering special cases comes from a rephrasing of the problem. Any set of n consecutive positive integers has the form $m, m+1, m+2, \ldots, m+(n-1)$ where m is a positive integer. Thus for a fixed positive integer n, we can test the statement in the special cases $m = 1$ and $m = 2$. Since $n!$ divides both $1 \cdot 2 \cdot \cdots \cdot n!$ and $2 \cdot 3 \cdot \cdots \cdot (n+1)$, the statement is seen to be true in these instances.

As Example 8 illustrates, the advice to examine special cases can be interpreted in more than one way in a given problem. Moreover, the dictum "examine special cases" is very broad and general. There is no single, unambiguous way of interpreting this guideline in every situation in which it is useful; and, as we saw in Example 8, there might be several distinct ways of forming special cases; some of these formulations might be helpful in solving the given problem and others might not. For more on the issue of examining special cases, see Chapter 3 of Schoenfeld's book [9].

A final word of warning is also in order. We are advising you to examine special cases so that you develop a better understanding of the problem. We are not suggesting that by considering a few particular cases, you will solve a problem or prove a theorem. In most instances you won't. Most mathematical statements describe properties of an infinite number of objects (e.g., the Goldbach conjecture) or a large finite number of objects. Thus by looking at a few special cases, you might find a counterexample to a general statement, some evidence for the validity of a statement, or an idea for a proof of a statement. But you cannot, for example, prove a statement about the behavior of elements in an infinite set by verifying the statement in a few special cases.

SIMPLIFY THE PROBLEM

Simplifying a problem usually amounts to making some assumptions about the problem that make it easier to handle. At this same time, these simplifying assumptions should not alter or significantly alter the essence of the problem. For instance, consider the following problem.

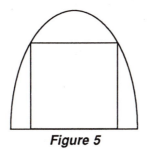

Figure 5

Example 9   The region in Figure 5 is bounded by a parabolic segment and a straight line segment that is perpendicular to the axis of the parabola. Suppose that the distance from the vertex of the parabola to the line segment is 4 units and that the length of the line segment is 4 units.

Find the dimensions of the largest rectangle that can be inscribed in the region.

To solve the problem, it is natural to look for a suitable representation. Perhaps the easiest way to represent the problem is to introduce a coordinate system and, by using the symmetry of the parabola, this representation can be made relatively simple. Place the region so that the line segment is horizontal and the midpoint of the segment is labeled 0. Label an arbitrary point on the horizontal segment x. The axis of symmetry of the parabola is therefore the line through 0 that is perpendicular to the line segment. Label an arbitrary point on this axis y. We have now introduced a coordinate system in which the given parabola has equation $y = 4 - x^2$ (Figure 6).

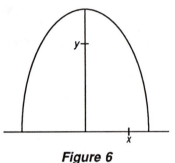

Figure 6

By using symmetry once again, we see that any rectangle inscribed in the region will have its lower vertices at points equidistant from 0, hence at points labeled x and $-x$ where $x > 0$. The area of such a rectangle is $2 \cdot x \cdot (4 - x^2)$ where $0 < x < 2$. Using techniques of calculus, we can solve for the value of x in the interval that yields the maximum area (Figure 7).

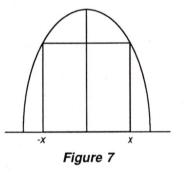

Figure 7

Often one can simplify a problem simply by choosing the units carefully. For example, in a problem in which lengths of time of millions or billions of years are considered, it is wise to let a single unit of time represent a million or a billion years.

EXERCISES §15

In exercises 1–7, identify the unknown. If appropriate, describe the unknown in two or more different ways, perhaps based on two or more distinct representations of the problem.

1. Among all pairs of positive numbers whose sum is 10, which pair has the largest product?
2. Among all pairs of positive numbers whose product is 10, which pair has the smallest sum?
3. Find all points at which the tangent line to the graph of $y = x^3 - x^2 - 2x + 1$ is horizontal.
4. Find all vectors in 3-space that are perpendicular to the vector $(1, 2, 3)$.
5. Find the point(s) of intersection of the circle $x^2 + 2x + y^2 = 1$ and the parabola $y = x^2 + x + 1$.
6. Find the area under the curve $y = \sin x$ (and above the x-axis) for $0 \leq x \leq \pi$.
7. Evaluate $\sum_{n=1}^{\infty} 1/n(n+1)$.

In exercises 8–15, develop an appropriate representation of the problem. The representation might be geometric or algebraic in nature. It should involve some suitable notation.

8. (a) Find all vectors in $\mathbf{R}^3$ perpendicular to both $(1, 3, -1)$ and $(0, 1, 2)$.
   (b) Find all vectors in $\mathbf{R}^4$ perpendicular to both $(1, 2, -1, 2)$ and $(0, 1, 2, 3)$.
9. Describe two or more ways of representing the following problem, known as the Cannibals-and-Missionaries Problem: Three cannibals and three missionaries are together on one side of a river. They wish to cross the river, and they have a boat that can carry two people. Describe a procedure for transporting all six people across the river in such a way that there are never more cannibals than missionaries on any one side at a given time.
10. Prove that in any group of six people there are either three mutual friends or three mutual strangers.
11. Prove $\sum_{n=1}^{\infty} 1/2^n = 1$.
12. Prove that if f and g are functions and $f(x) \geq g(x)$ for all x between a and b, then $\int_a^b f(x)\, dx \geq \int_a^b g(x)\, dx$.
13. Prove that if $0 \leq a_n \leq b_n$ for $n = 1, 2, 3, \ldots$, and the infinite series $\sum_1^{\infty} b_n$ converges, then $\sum_1^{\infty} a_n$ converges.
14. Prove that the diagonals of a parallelogram bisect each other.
15. (a) Describe two ways of representing a rational number.

(b) Describe two ways of distinguishing a rational number from an irrational number.

Section 16
ATTACKING
THE PROBLEM

The heart of the problem-solving process as presented by Polya and Schoenfeld lies in the second stage, Devising a Plan or Exploration. At this point the would-be problem solver possesses a firm understanding of the elements of the problem, can construct at least one representation of the problem, and perhaps can verify the result in special cases or can argue for the plausibility of a certain solution. In other words, the problem solver has a solid grip on the problem.

Even so, a solution might not be readily apparent. In fact with most substantive problems, solutions rarely appear at the understanding stage of the problem-solving process. What then should one do in order to find a solution?

Polya's advice is clear: Find a solvable related problem and use the related problem to obtain a solution to the original problem. Here is the basic strategy: given a problem P,

1. find a related problem P',
2. solve problem P',
3. use the solution to P' or the method of solution of P' to solve P.

This plan faces several apparent obstacles. First, how does one actually find a related problem? Second, the related problem itself might be difficult to solve; for example, it might be necessary to devise and solve a problem P'', related to P', and use the solution to P'' to solve P'. Finally, the transition from a solution to P' to a solution to P may be difficult to perform.

For the present we concentrate primarily on the first step of the plan outlined in the previous paragraph: How does one find a related problem? A general answer to this question is: You find a related problem by playing with the hypotheses and conclusion of the given problem. Not surprisingly, this general answer does not cover all cases. We give several examples illustrating the process of finding related problems. Our examples are organized under the three parts of phase II of Table 1.

ESSENTIALLY EQUIVALENT PROBLEMS

Example 1 We have already encountered one method of finding an essentially equivalent problem: To prove that an implication "If P then Q"

155

is true, consider the contrapositive, "If not-Q then not-P." The original implication holds if and only if the contrapositive holds. When we consider the contrapositive, we are replacing the original problem by a logically, hence essentially, equivalent problem. Here is another illustration of this method.

Problem: For any positive integer n, show that if $2^n - 1$ is a prime number, then n is a prime number.

First of all, let us clearly understand what we are trying to do. We want to establish an if-then statement. The nature of the conclusion "n is a prime number" suggests that an indirect proof might be in order, for if n is not prime, then n is expressible as a product of two smaller positive integers and this factorization of n gives us something tangible to work with. The contrapositive of the given statement is

For all $n \in \mathbf{Z}^+$, if n is not a prime number, then $2^n - 1$ is not a prime number.

Since n is not prime, there exist integers a and b such that $1 < a, b < n$ and $n = a \cdot b$. Thus $2^n - 1 = 2^{a \cdot b} - 1$.

Now what can we do with $2^{a \cdot b} - 1$? First, recall a property of exponents: $2^{a \cdot b} - 1 = (2^a)^b - 1$. The number $2^{a \cdot b} - 1 = (2^a)^b - 1$ is of the form $x^k - 1$ where $x = 2^a$ and $k = b$, and by the geometric sum property (see Example 2, Section 7, p. 63), for any real number x and any positive integer k,

$$x^k - 1 = (x - 1)(x^{k-1} + x^{k-2} + \cdots + x + 1).$$

Setting $x = 2^a$ and $k = b$ in this equation, we find that

$$2^n - 1 = 2^{a \cdot b} - 1 = (2^a)^b - 1 = (2^a - 1)((2^a)^{b-1} + \cdots + 2^a + 1).$$

Since $1 < a < n$, $1 < 2^a - 1 < 2^{a \cdot b} - 1 = 2^n - 1$. Therefore, $2^n - 1$ has a factor, namely $2^a - 1$, that is different from both 1 and itself, which means that $2^n - 1$ is not a prime.

Example 2   It is often very natural to replace a condition in a problem by an equivalent condition. For example, to show that a general linear system

$$ax + by = e$$

$$cx + dy = f$$

of two equations in two unknowns has a unique solution for any given e and f, it is useful to reformulate the problem in terms of matrices: Take the 2×2 matrix $A = \begin{bmatrix} a & b \\ c & d \end{bmatrix}$ and the column matrices $X = \begin{bmatrix} x \\ y \end{bmatrix}$ and $B = \begin{bmatrix} e \\ f \end{bmatrix}$. Then the given system of two linear equations in two unknowns x and y corresponds to the matrix equation

$$AX = B$$

where A and B are given matrices and X is an unknown matrix. For this matrix equation to have a solution for every B, it is necessary and sufficient to show that the matrix $A = \begin{bmatrix} a & b \\ c & d \end{bmatrix}$ is invertible. Furthermore, showing that A is invertible amounts to showing that $\det(A) = ad - bc \neq 0$. Thus, in two cases a condition has been replaced by an equivalent condition.

Another way to replace a problem by an essentially equivalent problem is to introduce so-called "auxiliary elements." By this we mean an object or feature that is added to the problem that, while not changing the problem, perhaps enhances or clarifies the relationships among the given elements of the problem. Auxiliary elements are frequently introduced in problems from algebra and geometry.

Example 3   Sketch the graph of the function $f(x) = \dfrac{x}{x - 1}$.

Before plunging ahead with the sketch, it is useful to play with the number $x/(x - 1)$ by adding 0 to the numerator in the form of $-1 + 1$:

$$\frac{x}{x - 1} = \frac{x + 0}{x - 1} = \frac{x - 1 + 1}{x - 1} = \frac{x - 1}{x - 1} + \frac{1}{x - 1} = 1 + \frac{1}{x - 1}.$$

Thus $f(x) = 1 + \dfrac{1}{x - 1}$. From this representation of the function, it is clear that the line $x = 1$ is a vertical asymptote and the line $y = 1$ is a horizontal asymptote for its graph. On the other hand, from the original representation, $f(x) = \dfrac{x}{x - 1}$, only the first conclusion (i.e., that $x = 1$ is a vertical asymptote for the graph) is clear.

Another method by which an auxiliary element can be introduced in algebra is to multiply by 1 in the form $1 = a/a$ where a is a suitable nonzero real number.

Example 4   In geometry, the introduction of an auxiliary element often provides a major insight into the solution of a problem or understanding of a theorem. Let us consider a proof of the Pythagorean Theorem. This world-famous theorem states that if a, b, and c are the lengths of the sides of a right triangle with c being the hypotenuse, then $c^2 = a^2 + b^2$.

Consider a typical right triangle labeled as follows:

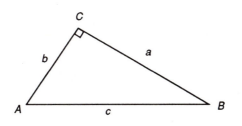

Since the statement of the Pythagorean Theorem describes a relationship between a, b, and c, it is reasonable to try to establish any connection whatever between a, b, and c. One way to relate a, b, and c is to introduce some similar triangles and this can be done by introducing an auxiliary element in the form of a perpendicular from C to side AB.

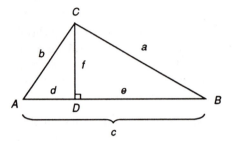

Since triangles ABC, ACD, and CBD are similar, $a/e = c/a$ and $b/d = c/b$ from which it follows that $a^2 + b^2 = ce + cd = c(d + e) = c^2$.

The method of introducing auxiliary elements is admittedly somewhat mysterious and tricky. How does one think of adding 0 in the form of $-1 + 1$? How does one think of dropping a perpendicular? The answers to these questions are not obvious. While the admonition, "introduce auxiliary elements" is quite general, the way in which one actually does the introducing depends on the specific domain of the problem. In algebra you often add 0 or multiply by 1. In geometry you draw a certain line or introduce a certain point. Perhaps the best general advice is to be aware of the possibility of introducing auxiliary elements and to develop ways of doing so that are geared to the specific areas of mathematics in which you are working.

The last technique of attacking a problem that falls under the general heading of considering an essentially equivalent problem is the method of *taking the problem as solved*. In this method you assume that you have a solution to the problem and try to deduce some properties of the solution that are known to be true. You then hope that you can reverse your steps: Starting with these true properties, you want to deduce the desired conclusion. This method is often used in algebra and geometry.

Example 5   Find the points of intersection of the circle C of radius 2 centered at $(1, 0)$ and the line L with slope 3 and y-intercept -7.

Solution.   Assume that (x_0, y_0) is a point lying on both C and L. Then (x_0, y_0) satisfies any and all equations that determine C and L. In particular, since (x_0, y_0) lies on C,

$$(x_0 - 1)^2 + y_0^2 = 2^2$$

and, since (x_0, y_0) lies on L,

$$y_0 = 3x_0 - 7.$$

Therefore, by substitution,

$$(x_0 - 1)^2 + (3x_0 - 7)^2 = 4$$

or

$$10x_0^2 - 44x_0 + 50 = 4$$

$$5x_0^2 - 22x_0 + 23 = 0.$$

By the quadratic formula

$$x_0 = \frac{22 \pm \sqrt{22^2 - 4 \cdot 5 \cdot 23}}{2} = 11 \pm \sqrt{6}.$$

The corresponding values of y_0 are $y_0 = 26 \pm 3\sqrt{6}$. Therefore, if (x_0, y_0) lies on the intersection of C and L, then either $(x_0, y_0) = (11 + \sqrt{6}, 26 + 3\sqrt{6})$ or $(x_0, y_0) = (11 - \sqrt{6}, 26 - 3\sqrt{6})$. One can now check that the converse also holds: If $(x_0, y_0) = (11 \pm \sqrt{6}, 26 \pm \sqrt{6})$, then (x_0, y_0) lies on both C and L.

SLIGHTLY MODIFIED PROBLEMS

The examples presented above illustrate how one can reformulate a given problem into an equivalent problem by recasting or reinterpreting parts of the original problem in ways that might appear to be different and yet are logically equivalent. We now provide illustrations of how one can modify a problem *slightly*. The usual procedure for doing this is to modify either the hypotheses or the conclusion in a substantial yet non-radical way. We present two examples.

The first example illustrates a standard method of problem modification—aiming for subgoals by relaxing and restoring conditions. A *subgoal* is a statement that lies "between" the hypothesis and conclusion of the given problem. If the original problem is an implication "If P then Q," then a subgoal is a statement R that the solver (1) hopes to derive from P and (2) intends to use to derive Q. Subgoals are often obtained by relaxing the conditions demanded in the conclusion. Our example comes from vector geometry.

Example 6 Find two unit vectors perpendicular to the vectors $\mathbf{v} = (1, 2, 3)$ and $\mathbf{w} = (1, -1, 4)$.

Solution. Our goal is to find two vectors, $\mathbf{u}_1$ and $\mathbf{u}_2$, such that $|\mathbf{u}_1| = |\mathbf{u}_2| = 1$ and $\mathbf{u}_1 \cdot \mathbf{v} = \mathbf{u}_2 \cdot \mathbf{v} = \mathbf{u}_1 \cdot \mathbf{w} = \mathbf{u}_2 \cdot \mathbf{w} = 0$. ($|\mathbf{v}|$ denotes the length of the vector $\mathbf{v}$. A unit vector is a vector $\mathbf{v}$ such that $|\mathbf{v}| = 1$.) A sketch of the situation should make it clear that the vectors $\mathbf{u}_1$ and $\mathbf{u}_2$ actually exist.

We aim for a somewhat less stringent goal. Subgoal (1): Find one unit vector perpendicular to $\mathbf{v}$ and $\mathbf{w}$. Thus we have created two problems from one:

(i) Find one unit vector perpendicular to **v** and **w**.

(ii) Using this unit vector, find two unit vectors perpendicular to **v** and **w**.

To solve (i) we again relax a requirement in order to obtain a subgoal. Subgoal (2): Find a nonzero vector perpendicular to **v** and **w**.

From the original problem we have constructed a chain of three problems: Given vectors $\mathbf{v} = (1, 2, 3)$ and $\mathbf{w} = (2, -1, 4)$,

(a) find a nonzero vector perpendicular to **v** and **w**,

(b) use the solution to (a) to find a unit vector perpendicular to **v** and **w**,

(c) use the solution to (b) to find two unit vectors perpendicular to **v** and **w**.

We leave as an exercise the task of solving problems (a)–(c).

Notice that in order to create the subgoals, we relaxed one of the conditions required of the solution to the problem: We ask for one unit vector, not two; we looked for a nonzero vector perpendicular to **v** and **w** rather than a unit vector perpendicular to **v** and **w**. Subgoals are usually created in this way. The restoration of the conditions amounts to moving from the subgoal to the goal. Also observe that the listing of subgoals provides a manageable and organized way of attacking the problem. In effect, an apparently long leap has been reduced to a sequence of relatively short leaps.

Example 7   The following problem illustrates how case analysis can be used to modfy a problem: Can one form a ten-digit integer by putting a digit between 0 and 9 in the empty boxes in the given table as follows: The digit in the box labeled 0 indicates the number of times 0 appears in the number, the digit in the box labeled 1 indicates the number of time 1 appears in the number, etc.?

0	1	2	3	4	5	6	7	8	9

This problem can be attacked by considering ten distinct cases, each determined by the digit that might be placed in the box labeled 0. For example, let Case 1 be that in which one tries to form the desired number by placing a 9 in box 0. It follows that the number contains nine 0's and must be 9000000000. But then the box labeled 9 has a 0 in it which means that the number does not contain a 9. This contradiction means that any number with the desired property cannot begin with a 9.

BROADLY MODIFIED PROBLEMS

Thus far our examples of related problems have in each case been closely connected with the original problems. We now look at examples in which the related problem is obtained by modifying the original problem significantly. These examples reveal two basic methods for creating broadly modified problems: analogy and generalization.

Analogy

Given a problem A, an analogous problem B has the same "structure" as A but takes place in a different setting. For example, A might be a problem in three-dimensional geometry involving lines and planes while B is a problem in two-dimensional geometry having to do with points and lines, or A might be a problem involving a function of three real variables while B is a problem involving a function of one real variable. We consider in detail an example from algebra.

Example 8   Show that if a, b, c, and d are real numbers strictly between 0 and 1, then $(1 - a)(1 - b)(1 - c)(1 - d) > 1 - a - b - c - d$.

Solution. Our first impulse might be to multiply out $(1 - a)(1 - b) \times (1 - c)(1 - d)$ and show that the resulting mess is greater than $1 - a - b - c - d$. However, the extent of the mess involved encourages us to seek another approach.

Let us look for a similar or analogous problem involving fewer variables. Consider the statement: If a and b are real numbers strictly between 0 and 1, then $(1 - a)(1 - b) > 1 - a - b$. This statement, while having the same form as the original, involves only two variables. Moreover, it is easy to check: Since $a \cdot b > 0$,

$$(1 - a)(1 - b) = 1 - a - b + a \cdot b > 1 - a - b.$$

We now add back a variable to obtain yet another analogous statement: If a, b, and c are real numbers strictly between 0 and 1, then

$$(1 - a)(1 - b)(1 - c) > 1 - a - b - c.$$

To prove this statement, perhaps we can use the result in the two-variable case considered above:

$$(1 - a)(1 - b) > 1 - a - b.$$

Multiplying both sides by $(1 - c)$, we have

$$(1 - a)(1 - b)(1 - c) > (1 - a - b)(1 - c)$$
$$= 1 - a - b - c + (a + b)c > 1 - a - b - c$$

since $(a + b)c > 0$.

Now what about the original problem? As can be checked, this argument extends to provide a proof that $(1 - a)(1 - b)(1 - c)(1 - d) > 1 - a - b - c - d$ if a, b, c, and d are strictly between 0 and 1. In fact, a pattern is emerging from these cases. As an exercise, try to describe that pattern and prove your description of it correct.

We mention in passing that several features of the Polya-Schoenfeld scheme are illustrated in the previous example. In studying the given problem, we found two analogous problems involving fewer variables. (See Table II.3(a), p. 142.) In answering the original problem, we were able to use the results of the analogous problems and the method used to solve these problems. Also each of the analogous problems can be regarded as extreme cases of the original problem: $c = d = 0$ for the first and $d = 0$ for the second (Table I.4(b), p. 142).

In many ways analogies are central to mathematics and mathematical thinking. Very often mathematicians find themselves working in a situation A that is reminiscent of a previously encountered situation B. In this case, B can be used as a model for the investigation of A. For instance, the study of functions of two or three variables is usually carried out in analogy with functions of one variable. Many concepts and theorems about functions of several variables are motivated by analogous concepts and theorems about functions of one variable. We will encounter more analogies later in this text. For now let us mention some important analogies in mathematics.

1. Functions of one variable: Functions of several variables:

    $y = f(x)$ $y = f(x_1, ..., x_n)$

2. Vectors in the plane: Vectors in 3-space:

    $v = (x, y)$ $v = (x, y, z)$

3. Sequences of real numbers: Functions of one variable:

    $a_1, a_2, \ldots, a_n, \ldots$ $y = f(x)$

4. Sums of real numbers: Integrals of functions:

    $a_1 + \cdots + a_n$ $y = \displaystyle\int_a^b f(x)\, dx$

5. Infinite series: Improper integrals:

    $\displaystyle\sum_{n=1}^{\infty} a_n$ $\displaystyle\int_1^{\infty} f(x)\, dx$

GENERALIZATION

A *generalization* of a given statement A is a statement B that includes A as a special case. In other words, the instances in which B applies include the instances in which A applies. Mathematicians are especially notorious for their tendency to generalize. One might reasonably question this practice. What are the advantages to generalization?

First, the process of generalizing often clarifies a problem. The essential features of a question are often obscured in a special case and are illuminated in the general situation. Details that seem important in a particular case are revealed to be less relevant when placed in a broader context. As a result a generalization of a statement is sometimes easier to handle than the original statement itself.

Second, through generalization we learn the limits of a given concept or method. By seeking to know the full extent to which a given statement is true, we develop a keener understanding of the concepts involved in it. If we can determine the instances in which a certain procedure applies, we can make maximum use of it while avoiding the pitfalls that arise from inappropriate use.

Finally, generalization can serve as a tool in research. Often a given idea can be generalized from a specific situation to a more encompassing setting in several ways. Each of these generalizations provides a starting point for further research and each such research effort creates a new body of mathematical knowledge. The process of generalization has the effect of extending and clarifying the concept that was initially considered.

To summarize in a few words, we generalize in order to know, to understand, to predict, and to create. The act of generalizing has one other notable benefit: It's fun.

Example 8 (revisited)   In working out this example, we observed that a certain type of inequality is valid in three distinct cases. Are these three cases (involving two, three, or four variables) part of a more general pattern? Suppose we allow any fixed finite number of real numbers in the inequality: If n is a positive integer greater than 1 and $a_1, ..., a_n$ are real numbers strictly between 0 and 1, then $(1 - a_1)(1 - a_2) \cdots (1 - a_n) > 1 - a_1 - a_2 - \cdots - a_n$. Note that when $n = 4$, the statement in Example 8 is recaptured.

Is this generalization true for all $n \geq 2$? Indeed it is for $n = 2, 3$, or 4. This suggests a proof by mathematical induction. For each integer $n \geq 2$, let $S(n)$ be the statement: If $a_1, \ldots, a_n$ are real numbers strictly between 0 and 1, then $(1 - a_1)(1 - a_2) \cdots (1 - a_n) > 1 - a_1 - a_2 - \cdots - a_n$.

To prove that $S(n)$ holds for $n \geq 2$, we must show that

(1) $S(2)$ is true;
(2) for any $m \geq 2$, if $S(m)$ is true, then $S(m + 1)$ is true.

We have already checked that $S(2)$ is true, so we consider the inductive step. We assume that $S(m)$ holds for some m and prove that $S(m + 1)$ is valid. We must prove that if $a_1, \ldots, a_{m+1}$ are real numbers between 0 and 1, then

$$(1 - a_1) \cdots (1 - a_{m+1}) > 1 - a_1 - \cdots - a_{m+1}.$$

From the inductive hypothesis we know that

$$(1 - a_1) \cdot \cdots \cdot (1 - a_m) > 1 - a_1 - \cdots - a_m.$$

Since $1 - a_{m+1} > 0$, we obtain from the latter inequality

$$(1 - a_1) \cdot \cdots \cdot (1 - a_m)(1 - a_{m+1}) > (1 - a_1 - \cdots - a_m)(1 - a_{m+1})$$

$$= 1 - a_1 - \cdots - a_m - a_{m+1}$$

$$+ (a_1 + \cdots + a_m) \cdot a_{m+1}$$

$$> 1 - a_1 - \cdots - a_m - a_{m+1}.$$

Thus the inductive step is established and $S(n)$ is true for all $n \geq 2$.

The following example comes from calculus.

Example 9 Recall the following theorem from one-variable calculus: If f is a differentiable function on an interval $[a, b] = \{x \mid a \leq x \leq b\}$ where a and b are fixed real numbers and if f achieves a relative maximum or a relative minimum at a point c where $a < c < b$, then $f'(c) = 0$. Problem: Generalize this statement to real-valued functions of two or more variables.

Let $f(x_1, \ldots, x_n)$ be a function of n variables. We seek a statement that relates the fact that f achieves a relative extremum (maximum or minimum) at a point to the vanishing of the "derivative" of f at that point. How can the notion of derivative be extended to functions of several variables? One method is through partial derivatives. Specifically, we take the differentiability of f to mean that the partial derivatives $\partial f/\partial x_1, \ldots, \partial f/\partial x_n$ all exist and are continuous. But where do they exist? To say precisely, we need to generalize the notion of interval to $\mathbf{R}^n$ and we have to define the concept of relative extremum of a function on $\mathbf{R}^n$. With these remarks in mind, we present a generalization of the one-variable theorem presented above.

Let R be a generalized rectangle in $\mathbf{R}^n$: $R = \{(x_1, \ldots, x_n) \mid a_1 < x_1 < b_1, \ldots, a_n < x_n < b_n\}$ where $a_1, \ldots, a_n, b_1, \ldots b_n$ are fixed real numbers. Suppose f is a real-valued function on R whose partial derivatives $\partial f/\partial x_1, \ldots, \partial f/\partial x_n$ exist and are continuous on R. If f has a relative maximum or relative minimum at a point c inside R, then $\partial f/\partial x_1(c) = 0, \ldots, \partial f/\partial x_n(c) = 0$.

Other generalizations are possible. For example, other kinds of regions can be considered. Also at least one other definition for the notion of differentiability is possible. Details can be found in textbooks on advanced calculus. In any case we have achieved our goal of generalizing the original statement about a function of one variable to a statement about functions of several variables. Question: Is the generalization true?

EXERCISES §16

In each of Exercises 1–7, find a related problem.

1. The only solution x, y, z, w to the linear system

$$x + 2y + w = 0$$

$$x - z + 2w = 0$$

$$y + z + 3w = 0$$

$$x + y + 2w = 0$$

is $x = y = z = w = 0$.

2. Find a unit vector u in the plane that makes an angle of $\pi/6$ with $(2, 3)$.

3. Prove that the infinite series $\sum_{n=1}^{\infty} 1/(n^2 + 3)$ converges.

4. The area of the ellipse $x^2/a^2 + y^2/b^2 = 1$ is πab.

5. For all real α, $\cos(\alpha/2) \cdot \cos(\alpha/4) \cdot \cos(\alpha/8) = (1/8)(\sin(\alpha)/\sin(\alpha/8))$.

6. (a) Evaluate $\sum_{n=1}^{\infty} 1/n(n + 1)$.

   (b) Evaluate $\sum_{n=1}^{\infty} n/(n + 1)!$.

7. Among all rectangular solids of surface area 20 sq. units, which has the largest volume?

In each exercise 8–12, give a reasonable generalization of the given statement.

8. Among all rectangles of perimeter 12, the square of side 3 has the largest area.

9. Among all pentagons (5-sided polygon) with a given perimeter, the regular pentagon (the one whose sides are of equal length) has the largest area.

10. Suppose 3000 points in the plane are given having the property that no three of the points are collinear. Prove that there exist 1000 disjoint triangles having the given points as vertices.

11. Any set of three vectors in the plane $\mathbf{R}^2$ is linearly dependent.

12. Let $A(r)$ and $C(r)$ be the area and circumference respectively of a circle of radius r. Then $dA/dr = C(r)$.

13. Notice that $1 \cdot 2 \cdot 3 \cdot 4 = 5^2 - 1$, $2 \cdot 3 \cdot 4 \cdot 5 = 11^2 - 1$. What is the general pattern?

Section 17
LOOKING BACK

Polya's admonition that we look back over our work has two thrusts. First, he encourages us to check our work by applying various tests to the solution—use of data, symmetry, and alternate derivations. Second, he points us toward future problems by having us try to derive new results from the result just established. In other words, looking back has both a backward and a forward orientation. In this section we concentrate on the first feature. However, we will comment briefly on the second and will illustrate both the backward and forward nature of checks throughout the text.

Polya divides the checks into two categories—specific tests and general tests. Our first three examples illustrate two of the specific tests while the next two examples illustrate general tests. In the first two examples, we show how to use reasonable estimates to check and to anticipate the solution to a problem.

Example 8 of Section 16 (revisited) In the inductive proof of the inequality

$$(1 - a_1)(1 - a_2) \cdots (1 - a_n) > 1 - a_1 - \cdots - a_n,$$

the fact that $0 < a_i < 1$ was used at a key point. A cruder check of the inequality follows from the observation that the left side, being a product of numbers between 0 and 1, is itself between 0 and 1, while the right side is clearly less than 1 and perhaps even less than 0. Thus the inequality does conform to reasonable estimates.

Example 9 of Section 15 (revisited) After suitably representing the original question, we are left with the problem of determining the rectangle of largest area whose base lies on the x-axis and whose upper vertices lie on the graph of $y = 4 - x^2$. Since the base and height of any such rectangle are at most 4, its area is at most 16. Thus the maximum area of an inscribed rectangle is somewhere between 0 and 16. Another estimate for the maximum area can be obtained by considering the rectangle whose lower vertices are at ± 1. The height of this rectangle is 3, hence its area is 6. Thus the area of the largest rectangle is somewhere between 6 and 16. As it happens, the largest rectangle has its lower vertices at $\pm\sqrt{4/3}$ and has area $32/3\sqrt{3} \approx 6.158$.

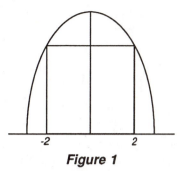

Figure 1

The next example shows how one can use symmetry to check (or to anticipate) an answer.

Example 1   Given a few minutes, one can multiply out

(∗) $(a + b + c)(a^2 + b^2 + c^2 + ab + ac + bc)$

to obtain

(∗∗) $a^3 + b^3 + c^3 + 2a^2b + 2a^2c + 2b^2a + 2b^2c + 2c^2a + 2c^2b + 3abc$

(Here we assume $a, b, c \in \mathbf{R}$. One way of checking the answer is through symmetry: each of the factors in (∗) is symmetric in a, b, and c. This means that if a and b (or a and c, or b and c) are interchanged in (∗), then the product stays the same. (For example, $(a^2 + b)(a + c)$ is not symmetric in a, b, and c.) Therefore, the expanded product must be symmetric in a, b, and c. The fact that the expression (∗∗) is symmetric in a, b, and c does not necessarily mean that (∗) and (∗∗) are equal, but it does provide some confirming evidence. Having confirming evidence of this sort is useful not only in the checking phase, but also when we are in the process of trying to discover a solution to a problem. For example, by symmetry it follows that the number of a^2b terms in the expanded product must be the same as the number of b^2a terms, etc.

What about some general tests that can be applied to a solution of a problem? In the next example we show how one can derive a result in a different way.

Example 2   The equality $1 + 3 + \cdots + (2n - 1) = n^2$ can be checked geometrically. Consider the squares

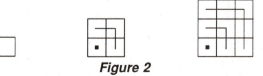

Figure 2

To obtain the 4×4 square from the 3×3 square, we glue three squares to the top of the 3×3 square, three squares to the right side, and one square to the upper right.

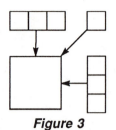

Figure 3

Thus $4^2 - 3^2 = 3 + 3 + 1 = 7$. In general the $(n + 1) \times (n + 1)$ square is obtained by adding $n + n + 1 = 2n + 1$ squares to the $n \times n$ square.

Notice that this check grows out of a different representation of the problem, namely a geometric representation. Thus, the looking back step can also be considered a step in the understanding phase of the problem-solving process. Indeed, the distinction between the two phases can be blurry. The realization that a given maneuver can be both an attempt to understand a problem and a device to check a solution underlies Polya's suggestion that looking back entails looking ahead.

Example 3 The Mean Value Theorem (MVT) is one of the most famous and important results in one-variable calculus. The statement of the MVT is the following:

Suppose f is a continuous function on the closed interval $[a, b]$. Suppose that $f'(x)$ exists for all x in the open interval (a, b). Then there exists at least one number $c \in (a, b)$ such that

$$f'(c) = \frac{f(b) - f(a)}{b - a}.$$

Even if you read a general proof of the MVT in a calculus book (or discover one for yourself), it is useful to verify the result in simple cases, for instance when f is a linear or quadratic function. For example, let us assume that $f(x) = x^2$. Then for all $a, b \in \mathbf{R}$ for which $a < b$,

$$\frac{f(b) - f(a)}{b - a} = \frac{b^2 - a^2}{b - a} = a + b.$$

Now $f'(x) = 2x$ for all $x \in \mathbf{R}$. Therefore

$$\frac{f(b) - f(a)}{b - a} = a + b = 2(\frac{a + b}{2}) = f'(\frac{a + b}{2})$$

and indeed $a < \dfrac{a + b}{2} < b$. Thus, if $f(x) = x^2$, then the statement of the MVT is satisfied for $c = \dfrac{a + b}{2}$.

By looking back and verifying the result of the MVT in this special case, we develop a more concrete understanding of the general statement. While the checking of the result in the special case is not, strictly speaking, necessary to our logical acceptance of the result, it does give us a strong feel for the statement.

Unit 4
DOING
MATHEMATICS

Section 18
CONTROLLING
YOUR THINKING

In the preceding units we have discussed the structure of mathematical proofs and some techniques for solving problems. In the process we have become aware of the general form of the arguments employed to prove theorems and the approaches that can be used to devise a proof of a theorem or a solution to a problem. Nonetheless, all of us, no matter how proficient we are at employing problem-solving strategies or at knowing proof methods, will occasionally become *stuck* on a problem. We meet what appears to be an insurmountable roadblock and we have no idea how to proceed. What do we do in these instances?

Perhaps the most useful general advice that can be offered is: STAY ACTIVE. Rather than waiting quietly for inspiration to strike, engage in steps that can help bring about a solution. Specifically, how does this advice translate into action? We offer a condensed list of suggestions, many of which have already been mentioned earlier in this book.

1. Read the problem or theorem several times. Rephrase the problem in alternative ways; for example, use quantifiers and connectives; use different but equivalent notation; or use different but equivalent definitions of the terms involved.

2. Play with the problem by drawing pictures (or forming other representations), by checking out examples, extreme cases, and other special cases, and by generalizing in various ways.

3. As a result of this active investigation, try to develop a problem-solving

169

or proof strategy.

4. Execute the chosen strategy and verify the result.

The moral is simple: Keep working, keep trying new things.

It might be apparent, however, that unharnessed activity is not sufficient. You cannot expect to try anything and to have it work. Discretion can be and often is appropriate. Thus, while you are actively working on a problem, it is wise to monitor your own activities and to control your actions.

For instance, suppose you find an approach that seems promising, but that after trying this approach for a while, you make no further progress. Then you should admit that perhaps your current strategy will not work and be willing to drop it in favor of some other. As another example, suppose you think of two possible ways of solving your problem. Before embarking on either, try to evaluate the relative advantages of the two methods. Perhaps one method will require significantly less work than the other and hence is preferable. As these remarks indicate, a diligent problem solver should be aware of more than the proof and problem-solving methods discussed in the previous two units. Successful problem solving also requires an awareness of one's actions, the flexibility to change actions, and the judgment to choose appropriate methods.

Specific examples that illustrate the need for control over one's actions are presented by Alan Schoenfeld in Chapter Four of his book, *Mathematical Problem Solving* [9]. Schoenfeld describes the efforts of several groups of college students and professional mathematicians to solve the following problem from geometry:

> Given a triangle T with fixed base B, show that it is always possible to construct with straightedge and compass a straight line that is parallel to B and divides T into two parts of equal area.

Each of the would-be problem solvers described by Schoenfeld possessed an adequate understanding of the problem and a knowledge of high school geometry sufficient to solve the problem.

Unfortunately, not all the problem solvers were successful. Those who failed did so because they persistently pursued blind alleys or simply wandered aimlessly around various aspects of the problem. Even though they were able to achieve some insights into the question, they did not reach a solution. The lack of success boiled down to their inability to use *efficiently* their mathematical resources and their knowledge of heuristics.

On the other hand, the successful problem solvers seemed to have firm control over their thought processes. They constantly took stock of their situation and made adjustments based on their findings. If the current course of action seemed to be leading nowhere, then that approach was abandoned or modified. Successful problem solvers frequently remind themselves of their goal and relate their progress on the problem to that goal. Throughout their analysis, efficient problem solvers ask themselves questions. This self-questioning helps the problem solver to monitor his or her position and to decide on appropriate steps.

The questions can take a variety of forms: Can I introduce an auxiliary element, in this case by dropping a perpendicular from a given vertex to the opposite side? Or: I seem to be stuck; what new approach can I try? Or: Have I used all the available information? Or: Now what is it exactly that I am trying to do? Again let us draw a moral: It is not just what you know, but how you use what you know.

In *Thinking Mathematically* [5], J. Mason *et al.* encourage readers to develop their own internal monitors and to become their own questioners. In this way, students of mathematics are forced to think not only about mathematical facts, but also about their own thought processes. They are encouraged to observe and to evaluate the current state of their understanding. Such activity has the benefit of leading not only to solutions of specific problems, but also to the development of general mathematical maturity.

To be your own questioner, you must be actively engaged in the problem. Such involvement can come from following the heuristic guidelines of Polya and can turn on such questions as: What is the problem? Can I draw a helpful picture? Can I find a useful specific case? Do I know of a similar problem? Have I seen this before? Have I used all the data? In addition to these heuristic questions, you can also ask yourself questions that are intended to help you assess and control your behavior. In particular, try to develop the habit of asking yourself the following three questions:

1. What do I want to do?
2. What am I doing?
3. Will what I am doing help me do what I want to do?

If you can answer questions 1 and 2 convincingly and can answer number 3 with a clear "yes," then proceed. Otherwise, stop and take stock of your situation.

A convincing answer to question 1 occurs only when the problem solver has a clear understanding of the problem. A convincing answer to question 2 occurs when the problem solver has a clear understanding of his or her own actions. An affirmative answer to question 3 results only when the actions of the problem solver are relevant to the goals of the problem.

Back to our original question: What do we do when we are stuck on a problem? Our answer has two related components: stay active and monitor your activities. Central to each of these components is the knack of asking questions. The questions might be specifically related to the given problem, might be of a heuristic nature, or might be aimed at monitoring your actions. In any case, as you work on a problem, keep talking to yourself by asking questions. The questions will help you keep tabs on yourself and may suggest approaches to the problem worth exploring. By actively and carefully searching for clues and insights, you will greatly increase your chances for finding a solution to the problem.

Section 19
ATTITUDES
AND
BELIEFS

The purpose of this section is to discuss the influence of attitudes and beliefs on mathematical performance. We will consider the general attitudes of American society toward mathematics and the beliefs of individuals about the nature of mathematics and about their own mathematical ability. This topic is considered in great depth by Alan Schoenfeld in Chapters 1 and 4 of *Mathematical Problem Solving* [9]. Interested readers are encouraged to consult that book for more details.

One of the most interesting things about the discipline of mathematics is what people think of it. There are many people who enjoy learning mathematics and solving mathematical problems. Most readers of this book fall into this category. Unfortunately, we seem to be in the vast minority of people. Judging from the popular press, comic strips, and random comments from ordinary people, most Americans either hate or strongly dislike mathematics. Negative attitudes toward mathematics permeate our culture and, in turn, influence the way people, especially school children, approach mathematics.

Many people wear their distaste for mathematics almost as a badge of honor. Parents will readily admit to their children that they never did like and never could do mathematics. For some reason, comments of this sort are accepted in our culture. How many parents would dare say to their children that they never did like and never could do reading or writing? Probably not very many. But somehow it is all right to admit to being a mathematical imbecile.

Mathematics-bashing is also a popular sport. One finds it in newspaper columns; at check-out counters in stores; on college campuses, even among so-called "enlightened" students; and in casual conversations among friends. Many people regard mathematics as a dull, boring, useless activity carried out by dull, boring, people wearing horn-rimmed glasses and pants with short cuffs.

The effect of these societal attitudes toward mathematics is to discourage students from doing well in math and to encourage them to stop their study of mathematics prematurely. The subtle message conveyed to students is: Math is not cool; don't bother with it. Female students are especially subject to this pressure, since, in addition, mathematics (and science in general) is not regarded as a feminine activity. Because a large segment of American society seems to undervalue mathematics, many capable students ultimately reject mathematics.

In addition to societal attitudes toward mathematics, each individual mathematics student carries his or her own beliefs about mathematics and about himself or herself as a mathematician. The beliefs held by a given individual greatly affect his or her mathematics performance. For many people, in fact, personal beliefs and attitude can be among the most important factors determining performance.

We will concentrate on negative opinions and beliefs about mathematics since such beliefs tend to lead to unsuccessful mathematical efforts. First, each individual brings beliefs about himself or herself with regard to the subject. For example, many students regard themselves as unable to do mathematics. This opinion can be due to experiences of failure or can be inherited from parents who hold and repeatedly utter the same belief. More interesting and more influential are the beliefs students have about mathematics itself.

Throughout their grade school years, students learn a number of algorithms or procedures for performing mathematical operations. For instance, they are taught how to add, subtract, multiply, and divide; how to solve simple systems of linear equations; and how to graph certain types of equations. As a result, students develop the belief that for every mathematical problem there is a specific procedure that produces a solution. In general, they come to believe that mathematics itself consists of procedures that apply to every problem or situation ever encountered.

What are some consequences of this belief? First, students who view mathematics as consisting only of procedures will reject aspects of mathematics such as proof, for example, that fail to be procedural. If there isn't a procedure for something, then it cannot be part of mathematics, so why bother with it? Such students can become very frustrated when faced with a problem that requires more than an algorithm for its solution. Often they end up pleading, "tell me how to do it," rather than trying to come up with a solution by applying their knowledge and reasoning ability.

A related belief, perhaps a corollary, is the notion that any mathematical problem that cannot be solved within five or ten minutes is impossible. Any student holding this opinion will quit on a problem rather readily. He or she will not experiment or "play with" the problem, nor will he or she carry the problem around for a day, mulling it over in order to find a solution. Such a student will also not be receptive to the problem-solving methods discussed in Unit 3. A student who stops working on a problem after ten minutes will be unable to solve problems of even moderate difficulty and therefore will eventually become discouraged about mathematics.

Another commonly held belief is that you have to be a genius to be good at mathematics. This belief is actually so widespread that it's virtually a cultural phenomenon. The consequences of this belief are rather dramatic. If I realize that I'm not a genius and believe that only geniuses can do math, then I should not expect to solve a difficult problem. Moreover, why should I even try working on anything more than a routine problem? Also, why should I try to understand the proof of a theorem, the justification of a geometric construction, or the derivation of a formula? Clearly, only a mathematical whiz can do these things. In effect, I am absolved of any responsibility with regard to the learning or the doing of mathematics. Obviously, any student who embraces this belief will have little incentive to perform well in mathematics.

Another interesting set of beliefs revolves around the role of proofs and formal reasoning in mathematics. Many students who have been through the standard pre-college high school mathematics curriculum of two years of algebra and a year each of geometry and pre-calculus believe that proofs belong only in geometry. Moreover, they believe that even in geometry, proofs are to be given only when called for by the teacher and have no relationship to other aspects of geometry such as the understanding of geometric constructions. The origins of this attitude toward proofs lie in the students' background. After all, proofs are emphasized more in geometry than in other high school mathematics courses. And proofs are often done in a very formal fashion and often seem to be divorced from the rest of mathematics. Thus it is on the basis of their high school experience that students form their beliefs concerning the role of proof in mathematics.

Any student who sees proofs as being restricted to geometry or as having no connection with the rest of mathematics will not regard proofs as necessary in justifying a given mathematical statement. For instance, such a student might justify empirically the statement "the sum of two odd integers is an even integer" by giving several examples: $3+7 = 10, 11+23 = 34$, etc. Therefore, the students might reason, "since the statement is true for several pairs of odd integers, then it must be true for all pairs of odd integers." As we have seen repeatedly, the gathering of empirical data is an important aspect of the mathematical enterprise. Many formulas and theorems came into existence as the result of specific calculations and the analysis of special cases. Nonetheless, a general theorem cannot be proved by being verified in a few special instances. A proof requires a general argument that accounts for every possible case.

So far we have concentrated on attitudes and beliefs that are antimathematical or negative in the sense that they tend to diminish mathematical performance. What about the other side of the coin? What are some positive attitudes toward mathematics? In particular, what are the attitudes of mathematicians as they do mathematics?

Most mathematicians know that any interesting problem will require time and thought before succumbing to a solution. Along the way, false starts will occur. There may be long periods in which no progress is made. In other words, mathematicians are not surprised when they are stuck on a problem. Nor are most mathematicians ashamed to admit to being stymied by a problem. They realize that being stuck is not synonymous with being stupid. Despite the disappointments and frustrations that can occur, a mathematician will continue to work actively on a problem, driven on by the challenge inherent in the problem and by the anticipation of the pleasure accompanying its solution.

In this unit we have followed the lead of Alan Schoenfeld and considered the influence of beliefs and attitudes on mathematical performance. As we have seen, many beliefs about mathematics held by many people are antimathematical in nature and bring about efforts that are unsuccessful. These unsuccessful experiences lead to frustrations and tend to reinforce negative attitudes. A vicious cycle ensues. Breaking out of this cycle can be difficult. It requires persistence, diligence, and confidence on the part of the individual involved. For readers of this book, a good way to proceed is to apply the problem-solving techniques discussed in Unit 3 while keeping in mind the control measures described in Section 18. Progress will occur— at times slowly, at times quickly. In time, your appreciation for, your understanding of, and your performance in mathematics will also increase. Good luck.

Let us conclude our excursion in the most natural way—by doing some mathematics. The following list of problems will allow you to apply many of the techniques discussed throughout this book. For some of the problems, the methods of solution are fairly straightforward. In other cases, they are more subtle and indirect. Many of the problems will require experimentation and conjecture. Some may well force you to try several approaches or points-of-view before an insight occurs. On some others, you may simply find yourself stuck.

No matter what the situation is with a given problem, though, it is hoped that as a whole, you will find the exercises interesting, challenging, and enjoyable. For the most part, the problems are *elementary*: They require little mathematical background beyond familiarity with the real number system and its basic properties, and the geometry of the Euclidean plane. (By the way, an "elementary" problem is not necessarily an easy problem. Some of the most famous open questions in mathematics such as the Fermat conjecture are elementary but difficult. Many great mathematicians have been stumped by elementary problems.) At the same time, the statement of a specific problem often reveals a fascinating relationship among the mathematical objects (be they integers, real numbers, or geometric figures) involved in the problem. For instance, the equality $1 + 2 + \cdots + n = n \cdot (n + 1)/2$, which is valid for every positive ineteger n, provides a non-obvious expression of an "open" sum, i.e., one involving n addends where n is indeterminate, in terms of a "closed" product involving only three factors. As we saw in Section 15, when this formula is interpreted geometrically, it becomes all the richer and more intriguing.

Perhaps our greatest hope is that you *enjoy* solving the problems, that the process of experimentation, investigation, understanding, conjecture, and verification lead not only to greater mathematical insight and maturity but also to pure intellectual pleasure.

1. Find the sum of the digits of all the numbers in the sequence $1, 2, 3, \ldots,$ 10000.

2. Consider the set, S, of all five-digit integers whose digits add up to 43. How many elements of S are divisible by 11? (There is a quick way of checking if an integer is divisible by 11: n is divisible by 11 if and only if the alternating sum of the digits of n starting with the ones digit is divisible by 11. For example, 12417 is not divisible by 11 since $7 - 1 + 4 - 2 + 1 = 9$ is not divisible by 11.)

3. An integer is called *palindromic* if it equals the integer formed by reversing the order of its digits. Thus 781187 is palindromic while 78178 is not. Is every four-digit palindromic integer divisible by 11?

4. Find a cubic polynomial whose roots are respectively twice as big as the roots of the cubic polynomial $P(x) = x^3 - 7x^2 + 5x + 2$.

5. Let $P(x) = a_n x^n + a_{n-1} x^{n-1} + \cdots + a_1 x + a_0$ be an arbitrary polynomial. Investigate the relationship between the roots of $P(x)$ and the polynomial $Q(x) = a_0 x^n + a_1 x^{n-1} + \cdots + a_{n-1} x + a_n$, defined by reversing the order of the coefficients of $P(x)$.

6. Let $a_1 = 10^{1/11}$, $a_2 = 10^{2/11}$, and in general $a_n = 10^{n/11}$. Find the least positive integer k such that the product $a_1 \cdot \cdots \cdot a_k$ exceeds 1000000.

7. For which integers n does the polynomial $x + 1$ divide the polynomial $x^n + 1$?

8. Let a and n be positive integers with $n > 1$. Prove that if $a^n - 1$ is a prime number, then $a = 2$. (Hint: remember the finite geometric sums; see p. 63.)

9. Given that $\sqrt{6}$ is irrational (see Section 5, Exercise 4(a)), prove that $\sqrt{2} + \sqrt{3}$ is irrational.

10. Let $\{f_n \mid n \geq 0\}$ be the *Fibonacci sequence*: $f_0 = f_1 = 1$ and, for $n \geq 2$, $f_n = f_{n-1} + f_{n-2}$.
    (a) Compute $f_1 f_4 - f_2 f_3$, $f_2 f_5 - f_3 f_4$, $f_3 f_6 - f_4 f_5$, and $f_4 f_7 - f_5 f_6$.
    (b) Make a conjecture based on the data in (a).

11. As in the previous problem, $\{f_n \mid n \geq 0\}$ is the Fibonacci sequence. Let $a_n = f_0 + f_1 + \cdots + f_n$.
    (a) Compute a_n for $n \leq 8$.
    (b) Make a conjecture about the value of a_n based on your data in (a).
    (c) Prove your conjecture.

12. Which terms in the Fibonacci sequence are even? Make a conjecture. Prove your conjecture.

13. Which terms of the Fibonacci sequence are divisible by 3? Make a conjecture. Prove your conjecture.

14. A $k \times k$ chessboard is called *defective* if exactly one square has been removed from the board. A *triomino* is an L-shaped tile that covers exactly three squares on a chessboard.

Prove that for each positive integer n, any $2^n \times 2^n$ defective chessboard can be tiled with triominoes.

15. Investigate the following question: How many squares are there on an $n \times n$ chessboard? For example, the 2×2 board contains 5 squares.

16. For each positive integer n let $d(n)$ denote the number of positive divisors of n.

   (a) Calculate $d(n)$ for $1 \leq n \leq 10$.
   (b) For which natural numbers n does $d(n) = 2$?
   (c) For which natural numbers n does $d(n) = 3$?
   (d) For which natural numbers n does $d(n) = 4$?
   (e) After conducting other experiments and investigation, record any other observations that you have made about $d(n)$.

17. At Polya High there are 1000 students and 1000 lockers (numbered 1–1000). At the start of the day, all the lockers are closed, then the first student comes by and opens every locker. Following the first student, the second student goes along and closes every second locker starting with locker number 2. The third student changes the state (if the locker is open, he closes it; if the locker is closed, he opens it) of every third locker starting with locker number 3. The fourth student changes the state of every fourth locker, etc., starting with locker number 4. Finally, the thousandth student changes the state of the thousandth locker. When the last student changes the state of the last locker, which lockers are open?

18. Investigate the following question: Which positive integers can be represented as a sum of any number of consecutive positive integers? For example, $3 = 2 + 1$ can be represented as such a sum but 2 apparently cannot be.

19. At a recent party attended by four couples, some people shook hands upon arrival. Of course no one shook his or her own hand or the hand of his or her spouse. Mr. Lewis, being a curious person, asked each of the other seven people how many hands did he or she shake and received seven *different* answers. How many hands did Mrs. Lewis shake? Consider analogous problems. Make a general conjecture. Prove your conjecture.

20. Suppose a and b are positive integers with $a \geq b$. Suppose the rules of basketball are modified to allow a points for a field goal and b points for a free throw. Investigate the following question: What are the possible scores attainable by a given team? (Assume no time limit on the length of a game.) Suggestions: Try special cases, e.g., $a = 3$, $b = 2$. Are there cases in which infinitely many scores are not attainable? Here is a somewhat related question: Which scores are attainable in American professional football? The scoring in pro football is: touchdown, 6

points; extra point, 1 point; field goal, 3 points; safety, 2 points.

21. Let f be a function on the collection $\mathbf{R}^+$ of positive real numbers such that $f(x) > 0$ for all x in $\mathbf{R}^+$ and

$$f(x + y) = \frac{f(x) \cdot f(y)}{f(x) + f(y)}$$

for all positive real numbers x and y.
   (a) Express $f(2 \cdot x)$ and $f(3 \cdot x)$ in terms of $f(x)$.
   (b) Conjecture a formula for $f(r \cdot x)$ that is valid for all positive rational numbers r and all positive real numbers x.
   (c) Prove your conjecture by a bootstrap proof.
   (d) Find a function f that has the properties listed above.

22. For any real numbers x and y, the *maximum* of x and y, denoted $\max(x, y)$, is defined by

$$\max(x, y) = \begin{cases} x & \text{if } x \geq y \\ y & \text{if } x < y. \end{cases}$$

   (a) Give a similar definition for the *minimum* of x and y.
   (b) Show that for all real x and y, $\max(x, y) = -\min(-x, -y)$.
   (c) Show that for all real x, y, and z, $\max(x, \min(y, z)) = \min(\max(x, y), \max(x, z))$.

23. Let A be a finite set with n elements where n is a positive integer.
   (a) How many subsets of A have exactly n elements?
   (b) How many subsets of A have exactly $n - 1$ elements?
   (c) How many subsets of A have exactly 2 elements?
   (d) How many subsets of A have exactly $n - 2$ elements?

24. For each positive integer n, what is the number of ways that the integers $1, 2, \ldots, n$ can be listed so that each integer after the first differs by 1 from an integer already listed? (For $n = 6$, 546321 is such a list while 231546 is not.)

25. How many two-element subsets of $\{1, \ldots, n\}$ do not contain consecutive integers?

26. Let $A(n)$ denote the number of ways that a positive integer n can be represented as a sum of 1s and 2s taking order into consideration. For example, $A(1) = 1$, $A(2) = 2$, and $A(3) = 3$.
   (a) Find $A(n)$ for $n \leq 6$.
   (b) Relate $A(n)$ to $A(n - 1)$ and $A(n - 2)$.

27. Consider an 8×8 chessboard with two opposite corners removed. Can such a board be covered by 31 dominoes, each of which covers exactly two squares on the chessboard?

28. Let $n \in \mathbf{Z}^+$. Suppose n lines are given in the plane in such a way that no two of the lines are parallel and no three meet at a point. Let p_n denote the number of regions into which the n lines divide the plane. Calculate p_n for $1 \leq n \leq 5$. Conjecture a general formula for p_n.

29. Let m and n be fixed positive integers. For $a, b \in \mathbf{R}$ define $a * b$ to be $(ma + nb)/(m + n)$.
    (a) Show that the binary operation $*$ on $\mathbf{R}$ has the *self-distributing property*: $a * (b * c) = (a * b) * (a * c)$.
    (b) Show that the arithmetic mean of a and b, $a * b = (a + b)/2$ has the self-distributive property.

30. Let A be a finite set. Show that the subsets of A (i.e., the elements of $P(A)$) can be listed in such a way that
    (a) $\emptyset$ is the first on the list,
    (b) each subset of A appears exactly once,
    (c) each subset on the list after the first is obtained either by deleting an element from the previous set on the list or by adding an element to the previous set on the list.

31. Suppose A and B are finite sets with m and n elements, respectively. Prove, or disprove and salvage: $A \cup B$ is a finite set with $m+n$ elements.

32. Prove, or disprove and salvage: Let A and B be sets. Then $P(A-B) \subseteq P(A) - P(B)$.

33. Let A, B, and C be arbitrary sets. Compare $(A-B) \cap C$ and $(A \cap C) - B$.

34. Prove that for all positive integers n, $n^3 - n$ is divisible by 3. Now give another proof of this statement.

35. Prove that for all positive integers n, $n^5 - n$ is divisible by 5.

36. Investigate the following assertion: The product of any four consecutive integers is always one less than a perfect square.

37. (a) Calculate $1 - 3$, $1 - 3 + 5$, $1 - 3 + 5 - 7$, $1 - 3 + 5 - 7 + 9$, etc.
    (b) State a general conjecture based on the data in (a).
    (c) Prove your conjecture.

38. (a) Calculate $1 - 4$, $1 - 4 + 9$, $1 - 4 + 9 - 16$, etc.
    (b) State a general conjecture based on the data in (a).
    (c) Prove your conjecture.

39. The *Lucas sequence* $\{L_n \mid n \geq 0\}$ is defined by the rules $L_0 = 1$, $L_1 = 3$ and for $n \geq 2$, $L_n = L_{n-1} + L_{n-2}$. (Note that all terms beyond the first two are obtained by adding the prevous two terms, just as with the Fibonacci sequence. The difference between the sequences is that the Lucas sequence begins with 1 and 3, while the Fibonacci starts with 1 and 1.)
    (a) Calculate $L_0 L_3 - L_1 L_2$, $L_1 L_4 - L_2 L_3$, and $L_2 L_5 - L_3 L_4$.
    (b) State a general conjecture based on the data in (a).
    (c) Prove your conjecture.

40. Prove that for each positive integer n, there exists an integer k such that $3^k \leq n < 3^{k+1}$.

41. A *base 3 expansion* of a positive integer n is a representation of n in the form $n = a_0 + a_1 \cdot 3 + a_2 \cdot 3^2 + \cdots + a_k \cdot 3^k$ where each a_i is either 0, 1, or 2.
    (a) Find base 3 expansions for 88 and 231.
    (b) Prove: Every positive integer has a base 3 expansion.

(c) (Challenge) Prove that every positive integer has a unique base 3 expansion. Specifically show that for every positive integer n, if $n = a_0 + a_1 \cdot 3 + \cdots + a_k \cdot 3^k$ and $n = b_0 + b_1 \cdot 3 + \cdots + b_m \cdot 3^m$ where each a_i is 0, 1, or 2 and each b_i is 0, 1, or 2, then $k = m$, and $a_0 = b_0$, $a_1 = b_1$, $a_2 = b_2$, ..., $a_k = b_k$.

42. Let b be any integer greater than 1. Define the concept of a *base b expansion* of a positive integer n.

43. Prove that for every positive integer n there exists a positive integer k such that $k! \le n < (k+1)!$.

44. Let n be a positive integer. A *factorial-base expansion of n* is an expression of the form $n = a_1 + a_2 \cdot 2! + a_3 \cdot 3! + \cdots + a_k \cdot k!$ where for $1 \le i \le k$, a_i is an integer and $0 \le a_i \le i$.
   (a) Find factorial-base expansions of 105 and 478.
   (b) Prove that every positive integer has a factorial-base expansion.
   (c) Prove that every positive integer has a unique factorial-base expansion.

45. Again let $\{f_n \mid n \ge 0\}$ denote the Fibonacci sequence.
   (a) Evaluate $f_1 + f_3 + \cdots + f_{2n-1}$ for $1 \le n \le 5$.
   (b) Conjecture and prove a general formula for the sum in (a).

46. Repeat Problem 45 for the sum $f_1^2 + f_2^2 + \cdots + f_n^2$.

47. Establish the identity $1/(f_{n-1} \cdot f_{n+1}) = 1/(f_{n-1} \cdot f_n) - 1/(f_n \cdot f_{n+1})$.

48. Evaluate the infinite series $\displaystyle\sum_{n=1}^{\infty} 1/(f_{n-1} \cdot f_{n+1})$.

49. Evaluate the infinite series $\displaystyle\sum_{n=1}^{\infty} f_n/(f_{n-1} \cdot f_{n+1})$.

50. Redo each of Problems 45–49 using the Lucas sequence (see Problem 39) in place of the Fibonacci sequence.

51. Let A and B be nonempty sets. Prove that if $A \times B$ is a finite set, then A and B are finite sets.

52. Let $f : A \to B$ be a function. For $x, y \in A$, define $x \sim_f y$ if $f(x) = f(y)$.
   (a) Prove: $\sim_f$ is an equivalence relation on A.
   (b) If f is injective, how large is each equivalence class?

53. In a round-robin tennis tournament, every player plays every other player exactly once. Assume that in each match there is a winner and a loser, i.e., no ties occur. Define a *top* player to be one who, for every other x, either beats x or beats a player who beats x.
   (a) Prove that for every positive integer n, every n-player tournament has a top player.
   (b) Give an example of an n-player tournament with more than one top player.

54. Let $\{p_n \mid n \geq 2\}$ be the sequence defined as follows: $p_2 = \dfrac{2^3 - 1}{2^3 + 1}$, $p_3 = \left(\dfrac{2^3 - 1}{2^3 + 1}\right)\left(\dfrac{3^3 - 1}{3^3 + 1}\right)$, $\ldots, p_n = \left(\dfrac{2^3 - 1}{2^3 + 1}\right)\left(\dfrac{3^3 - 1}{3^3 + 1}\right) \cdot \ldots \cdot \left(\dfrac{n^3 - 1}{n^3 + 1}\right)$.

(In other words, for $n \geq 3$, $p_n = p_{n-1} \cdot \left(\dfrac{n^3 - 1}{n^3 + 1}\right)$.)

   (a) Evaluate p_n for $2 \leq n \leq 10$.
   (b) Write p_n as a fraction and try to notice patterns in the sequence of numerators and the sequence of denominators. (Hint: for some values of n, it might be useful not to write p_n in lowest terms.)
   (c) Conjecture a formula for p_n. (Hint: both the numerator and denominator can be expressed as quadratic functions of n.)
   (d) Prove your conjecture.
   (e) Show that $\lim\limits_{n \to \infty} p_n = 2/3$.

55. (Another approach to the previous problem.) Let $\{p_n \mid n \geq 2\}$ be defined as in Problem 54.
   (a) Show that $n^3 - 1 = (n-1)(n^2+n+1)$ and $n^3+1 = (n+1)(n^2-n+1)$.
   (b) Use part (a) to show that $p_n = \dfrac{2}{3} \dfrac{n^2 + n + 1}{n(n + 1)}$.
   (c) Use part (b) to show that $\lim_{n \to \infty} p_n = 2/3$.

56. Let $f : A \to B$ be a function and let R be an equivalence relation on B. Define a relation S on A as $x \, S \, y$ if $f(x) \, R \, f(y)$. Prove: S is an equivalence relation on A. Give an example of an equivalence relation S on a set A that arises in this way.

57. Prove that for each positive integer n, the fraction $\dfrac{21n + 4}{14n + 3}$ is in lowest terms. In other words, prove that if d is a positive integer that divides both $21n + 4$ and $14n + 3$, then $d = 1$.

58. Express L_n, the nth term of the Lucas sequence, as a sum of some terms of the Fibonacci sequence.

59. Prove the arithmetic-geometric inequality: Let a_1 and a_2 be positive real numbers. Then $\sqrt{a_1 a_2} \leq \dfrac{a_1 + a_2}{2}$. (Note: the number $\sqrt{a_1 a_2}$ is called the *geometric mean of a_1 and a_2* while $\dfrac{a_1 + a_2}{2}$ is called the *arithmetic mean of a_1 and a_2*.)

60. Show that the function $f : \mathbf{N} \times \mathbf{N} \to \mathbf{Z}^+$ defined by $f(a, b) = 2^a(2b+1)$ is bijective.

61. (Geometric interpretation of the arithmetic-geometric inequality.) Consider a line segment of length $a_1 + a_2$ with endpoints A and B. Let C be the point on the segment such that the length of AC is a_1 and the length of CB is a_2. Consider next a circle with diameter AB. Let DC be the half chord of the circle that is perpendicular to AB at C.
   (a) Relate the lengths AD and BD to $a_1 + a_2$.
   (b) Find the length CD.
   (c) Interpret the arithmetic-geometric inequality geometrically.

62. Symmetric difference. Let A and B be subsets of a set U. The *symmetric difference of A and B*, written $A + B$, is the set $A + B = (A - B) \cup (B - A)$. (Note: Some authors write $A \Delta B$ in place of $A + B$.)
   (a) Let $A = \{1, 2, 3\}$, $B = \{3, 4, 5\}$, and $C = \{4, 5, 6\}$. Evaluate $A + B$, $A + C$, $B + C$, $(A + B) + C$, and $A + (B + C)$.
   (b) Draw a Venn diagram for $A + B$.
   (c) Prove: $A + B = (A \cup B) - (A \cap B)$.
   (d) Prove: $+$ is commutative.
   (e) Prove: For any $A \subseteq U$, $A + A = \emptyset$ and $A + \emptyset = A$.
   (f) Prove: For all $A, B \in P(U)$, $A^c + B^c = A + B$.
   (g) What is $A + A^c$ for an arbitrary set $A \subseteq U$?
   (h) Draw a Venn diagram to convince yourself that $+$ is associative.
   (i) Suppose U is finite. Describe the binary sequence of $A + B$ in terms of the binary sequences of A and B.

63. Show that each integer in the sequence 49, 4489, 444889, 44448889, ... is a perfect square.

64. For $n \geq 1$, let $a_n = \dfrac{1}{1 \cdot 4} + \dfrac{1}{4 \cdot 7} + \cdots + \dfrac{1}{(3n - 2)(3n + 1)}$.
   (a) Evaluate a_n for $1 \leq n \leq 8$.
   (b) Conjecture a general formula for a_n.
   (c) Prove your conjecture.

65. For $n \geq 1$, let $b_n = \dfrac{1}{1 \cdot 5} + \dfrac{1}{5 \cdot 9} + \cdots + \dfrac{1}{(4n - 3)(4n + 1)}$.
   (a) Evaluate b_n for $1 \leq n \leq 8$.
   (b) Conjecture a general formula for b_n.
   (c) Prove your conjecture.

66. State and prove a general theorem that includes the results of Problems 64 and 65 as special cases. Note that your general theorem should also include the results of p. 70, Exercises 11 and 12 as special cases.

67. Let A be a set with n elements.
   (a) How many subsets of A contain an even number of elements? Experiment with small values of n before making a conjecture.
   (b) Prove your conjecture.

68. Let $F(\mathbf{R})$ denote the set of functions from $\mathbf{R}$ to $\mathbf{R}$. For $f, g \in F(\mathbf{R})$, we say that $f \sim g$ if $f - g$ is a differentiable function (i.e., $f - g$ has a derivative at each $x \in \mathbf{R}$). Show: $\sim$ is an equivalence relation on $F(\mathbf{R})$.

69. State a general theorem that includes as special cases the formulas

$$\frac{1}{1 \cdot 2} + \frac{1}{2 \cdot 3} + \cdots + \frac{1}{n(n + 1)} = \frac{n}{n + 1}$$

and $\quad \dfrac{1}{2 \cdot 3} + \dfrac{1}{3 \cdot 4} + \cdots + \dfrac{1}{(n + 1)(n + 2)} = \dfrac{n}{2(n + 2)}$.

70. Let $a_n = \dfrac{1^2}{1 \cdot 3} + \dfrac{2^2}{3 \cdot 5} + \cdots + \dfrac{n^2}{(2n-1)(2n+1)}$.

    (a) Evaluate a_n for $a \le n \le 6$.

    (b) Conjecture and prove a formula for a_n.

71. Suppose $f : A \to B$ and $g : B \to A$ are functions such that $g \circ f = I_A$ and $f \circ g = I_B$. Show: f and g are both bijections and that $f = g^{-1}$. What is f^{-1}?

72. (a) Check that $10! = 7!6!$.

    (b) Prove that there exist infinitely many integers a, b, and c, all greater than 1, such that $a! = b!c!$.

73. Define a relation R on $\mathbf{R}^2$ as follows: For $(x_1, y_1), (x_2, y_2) \in \mathbf{R}^2$, $(x_1, y_1) \, R \, (x_2, y_2)$ if and only if $|y_1 - x_1| = |y_2 - x_2|$.

    (a) Show that R is an equivalence relation on $\mathbf{R}^2$.

    (b) Describe the equivalence classes geometrically. In particular, describe the equivalence classes of $(2, 5)$ and $(1, -1)$.

74. Let a and b be fixed positive real numbers. Determine $\lim\limits_{n \to \infty} (a^n + b^n)^{1/n}$.

    (Hint: You may use the fact that $\lim\limits_{n \to \infty} c^{1/n} = 1$ for any fixed positive real c.)

75. Arrange the numbers $1, 2, \ldots, n$ consecutively about the circumference of a circle. Now, remove number 2 and proceed in order removing every other number among those that remain until only one number is left. Let $f(n)$ denote the final number that remains. For example, $f(2) = 1$, $f(3) = 3$, $f(4) = 1$, and $f(5) = 3$.

    (a) Show that $f(2n) = 2f(n) - 1$.

    (b) Show that $f(2n + 1) = 2f(n) + 1$.

    (c) Conjecture a formula for $f(n)$.

76. Suppose 3000 points in the plane are given with the property that no three of the points are collinear. Prove that there exist 1000 disjoint triangles having the given points as vertices.

Hints and Solutions to Selected Exercises

Section 1

1. (a)

P	Q	P ⇒ Q	P ⇒ (P ⇒ Q)
T	T	T	T
T	F	F	F
F	T	T	T
F	F	T	T

(c)

P	Q	R	not-(Q and R)	P ⇒ (not-(Q and R))
T	T	T	F	F
T	T	F	T	T
T	F	T	T	T
T	F	F	T	T
F	T	T	F	T
F	F	T	T	T
F	F	F	T	T

(f)

P	Q	P ⇒ Q	not-P ⇒ not-Q	(P ⇒ Q) ⇒ (not-P ⇒ not-Q)
T	T	T	T	T
T	F	F	T	T
F	T	T	F	F
F	F	T	T	T

3. (a) From the following partial truth table,

P	Q	P ⇒ Q
T	T	T
T	F	F

it follows that Q must be true.

6. (a)

P	Q	not-(P ⇒ Q)	P and (not-Q)
T	T	F	F
T	F	T	T
F	T	F	F
F	F	F	F

9.

P	Q	P ⇔ Q	A = (P ∧ Q)∨ ((not-P) ∧ (not-Q))	(P ⇔ Q) ⇔ A
T	T	T	T	T
T	F	F	F	T
F	T	F	F	T
F	F	T	T	T

12. (a)

P	Q	$(P \vee Q)$	$(P \wedge \sim Q)$	$(P \vee Q) \Rightarrow (P \wedge \sim Q)$
T	T	T	F	F
T	F	T	T	T
F	T	T	F	F
F	F	F	F	T

(b) Try $\sim Q$.

14. (a)

P	$P \triangledown P$	not-P
T	F	F
F	T	T

(b) $P \vee Q$ is logically equivalent to $(P \triangledown Q) \triangledown (P \triangledown Q)$.

Section 2

1. (b) $\{4 \pm \sqrt{11}\}$ (c) $\{\pm 1/\sqrt{2}\}$

2. (a) $\{1, 2, 5\} = \{5, 2, 1\}$ since the two sets have precisely the same elements.

   (d) $\{\{a\}, \{\{a\}\}\} \neq \{a\}$ since $\{\{a\}\} \in \{\{a\}, \{\{a\}\}\}$, but $\{\{a\}\} \notin \{\{a\}\}$.

3. (a) $a = -1$, $b = 2$

6. (a) 00000   (c) 11100

7. (a) $\emptyset - 00$
           $\{1\} - 10$
           $\{2\} - 01$
           $\{1, 2\} - 11$

Section 3

1. (a) Contrapositive: For all real numbers a and b, if $a \cdot b \not> 0$, then $a \not> 0$ or $b \not> 0$.
   (b) Converse: For all real numbers a and b, if $a \cdot b > 0$, then $a > 0$ or $b > 0$. The converse is false since $(-1) \cdot (-1) > 0$ yet $-1 \not> 0$.

4. (a) Contrapositive: If there exists x in (a, b) such that $f'(x) \neq 0$, then f is not constant on $[a, b]$.

11. (a) Contrapositive: If the angles of triangle T are not equal, then the sides of T are not equal.

    (b) Converse: If T is an equiangular triangle, then T is an equilateral triangle. The converse is true.

12. (a) There exists x such that x is an integer and x is a perfect square.
    (b) Negation: For all integers x, x is not a perfect square.

16. (a) For all real numbers x, if x is irrational, then $\sqrt{x}$ is irrational.

(b) There exists a number x such that x is irrational and $\sqrt{x}$ is rational.

19. (a) For all x and y, if x and y are irrational, then $x + y$ is irrational.

21. (a) $(\exists x)(\text{not-}P(x) \wedge \text{not-}Q(x))$.

(d) $(\forall x)(\text{not-}P(x) \vee \text{not-}Q(x))$.

Section 4

2. Let x and y be integers. Suppose that x is even. Then there exists an integer n such that $x = 2 \cdot n$. Thus $x \cdot y = (2 \cdot n)y = 2 \cdot (n \cdot y)$ and $x \cdot y$ is an even integer. The argument is similar in case y is even.

6. Given that $b = a \cdot n$ for some integer n, $b^2 = (a \cdot n)^2 = a^2 \cdot n^2$; hence a^2 is a factor of b^2.

Section 5

2. The contrapositive reads: If a is even or b is even, then $a \cdot b$ is even. A proof can be patterned after the argument in Example 1.

3. Suppose $\sqrt{8}$ is rational. Then there exist integers a and b such that $b \neq 0$ and $\sqrt{8} = a/b$. By properties of square roots, $\sqrt{8} = \sqrt{4} \cdot \sqrt{2} = 2\sqrt{2}$, and hence $\sqrt{2} = a/(2 \cdot b)$ is a rational number. This conclusion, however, contradicts Example 4. Therefore, $\sqrt{8}$ is irrational.

6. Suppose $\log_2(5) = a/b$ where a and b are integers. We can assume that $a > 0$ and $b > 0$. Then $2^{a/b} = 5$, which implies that $2^a = 5^b$. Thus $2^a = 5^b$ is an integer that is both even and odd. This contradiction implies that $\log_2(5)$ is irrational.

Section 6

1. By Axiom 9, for x in $\mathbf{R}$, $x \neq 0$, there exists at least one element y in $\mathbf{R}$ such that $x \cdot y = y \cdot x = 1$. Suppose there exists another element y' in $\mathbf{R}$ such that $x \cdot y' = y' \cdot x = 1$. Since $x \cdot y = 1$, $y' \cdot (x \cdot y) = y' \cdot 1 = y'$. But by associativity, $y' = y' \cdot (x \cdot y) = (y' \cdot x) \cdot y = 1 \cdot y = y$. Thus there exists exactly one element of $\mathbf{R}$ having the desired property.

2. (a) Let x be in $\mathbf{R}$. Then $x \cdot 0 = x \cdot (0 + 0) = x \cdot 0 + x \cdot 0$ by Axioms 3 and 11. By Axiom 4, there exists y in $\mathbf{R}$ such that $x \cdot 0 + y = 0$. Therefore, $0 = x \cdot 0 + y = (x \cdot 0 + x \cdot 0) + y = x \cdot 0 + (x \cdot 0 + y) = x \cdot 0 + 0 = x \cdot 0$ by Axioms 2 and 3.

3. (a) By 2(d) and (b), $(-1) \cdot (-x) = -(-x) = x$.

6. (a) If x and y are in $\mathbf{R}$ and $x < y$, then $y - x$ is in $\mathbf{R}^+$. We show that $(-x) - (-y)$ is in $\mathbf{R}^+$:

$$
\begin{aligned}
y - x = y + (-x) &= (-x) + y & \text{by Axiom 5} \\
&= (-x) - (-1) \cdot y & \text{by Axiom 3(b)} \\
&= (-x) + (-1) \cdot (-y) & \text{by Axiom 2(c)} \\
&= (-x) - (-y) & \text{by Axiom 2(c)}
\end{aligned}
$$

Therefore, $(-x) - (-y)$ is in $\mathbf{R}^+$ and hence $(-x) > (-y)$.

7. (a) By Axiom 13, either 1 is in $\mathbf{R}^+$ or -1 is in $\mathbf{R}^+$; however, it is not the case that both 1 and -1 are in $\mathbf{R}^+$. Suppose 1 is not in $\mathbf{R}^+$. Then (-1) must be in $\mathbf{R}^+$. But by Axiom 12, it follows that $(-1) \cdot (-1) = 1$ is in $\mathbf{R}^+$, which is impossible since 1 and -1 cannot both be in $\mathbf{R}^+$.

11. (a) Clearly 5 is an upper bound for the collection A of all real numbers less than 5. Suppose $x < 5$. Then $x < x + (5 - x)/2$ and $x + (5 - x)/2 = (5 + x)/2 < 5$. Thus $x + (5 - x)/2$ is in A and is greater than x. Therefore x is not an upper bound for A. Thus no real number less than 5 is an upper bound for A. Therefore, 5 is the least upper bound for A.

(b) Let B be the collection of all real numbers x such that $0 < x < 1$. Then B is bounded above in A by $3/2$. Let x be an upper bound for A. Then $x > 1$, for otherwise $x < x + (1 - x)/2$, which is in B. But $1 < y = 1 + (x - 1)/2 < x$. Thus y is an upper bound in A for B and $y < x$. Therefore, no least upper bound for A exists in B.

14. (a) If $a > b$, then $a - b > 0$. Since $c > 0$, $(a - b) \cdot c > 0$ by Axiom 12. Thus $(a - b) \cdot c = ac - bc > 0$ and $ac > bc$.

Section 7

3. For each positive integer n, define $S(n)$: For each real $x \geq -1$, $\quad (1 + x)^n \geq 1 + nx$. $S(1)$ holds since $1 + x \geq 1 + x$ for all $x \geq 1$. Suppose $S(n)$ holds and consider $(1 + x)^{n+1}$. By inductive hypothesis, $(1 + x)^{n+1} = (1 + x)^n (1 + x) \geq (1 + nx)(1 + x) = 1 + (n + 1)x + nx^2 \geq 1 + (n + 1)x$ since $nx^2 \geq 0$. Therefore by PMI, $S(n)$ holds for all $n \geq 1$.

4. For each positive integer n, define $S(n)$: $1^2 + \cdots + (2n - 1)^2 = (4n^3 - n)/3$. $S(1)$ holds since $1^2 = 1 = (4 \cdot 1^3 - 1)/3$. Suppose $S(n)$ holds. Then $1^2 + \cdots + (2(n + 1) - 1)^2 = (4n^3 - n)/3 + (2n + 1)^2 = (4n^3 - n + 12n^2 + 12n + 3)/3 = (4(n + 1)^3 - (n + 1))/3$. Therefore, $S(n)$ implies $S(n + 1)$, and by PMI, $S(n)$ holds for all $n \geq 1$.

9. For $n \geq 2$, define $S(n)$: If $a_1, \ldots, a_n$ are odd integers, then $a_1 \cdot \cdots \cdot a_n$ is an odd integer. $S(2)$ holds by Example 1 of Section 4, p. 38. Suppose

$S(n)$ holds and let $a_1, \ldots, a_{n+1}$ be odd integers. Then $a_1 \cdots \cdots a_n$ is an odd integer by inductive hypothesis. Since $S(2)$ holds, $(a_1 \cdots \cdots a_n) \cdot a_{n+1} = a_1 \cdots \cdots a_{n+1}$ is an odd integer. Therefore by PMI the statement $S(n)$ holds for $n \geq 2$.

11. (a) $a_1 = 1/2$, $a_2 = 2/3$, $a_3 = 3/4$, $a_4 = 4/5$, $a_5 = 5/6$.
    (b) Conjecture: For $n \geq 1$, $a_n = n/(n+1)$.
    (c) The inductive proof of this conjecture is straightforward.
    (d) $\sum\limits_{k=1}^{\infty} 1/k(k+1) = 1$ since, for $n \geq 1$, the nth partial sum of this series is $a_n = 1/1{\cdot}2 + \cdots + 1/n(n+1) = n/(n+1)$ and $\lim\limits_{n \to \infty} (n/(n+1)) = 1$.

13. (a) $c_1 = 1 = 1^2$, $c_2 = 9 = 3^2$, $c_3 = 36 = 6^2$, $c_4 = 100 = 10^2$, $c_5 = 225 = 15^2$.
    (b) Note that $1 = 1$, $3 = 1+2$, $6 = 1+2+3$, $10 = 1+2+3+4$ and $15 = 1+2+3+4+5$. Thus we conjecture: For $n \geq 1$, $c_n = (1 + \cdots + n)^2 = (n(n+1)/2)^2$. Again, the inductive proof of this conjecture is straightforward.

19. (a) $m_2 = 3$, $m_3 = 7$, $m_4 = 43$.
    (b) For $n \geq 1$, define $S(n)$: $m_{n+1} = m_1 \cdots \cdots m_n + 1$. Inductive step: $m_1 \cdots \cdots m_{n+1} + 1 = (m_1 \cdots \cdots m_n) \cdot m_{n+1} + 1 = (m_{n+1} - 1) \cdot m_{n+1} + 1$ by inductive hypothesis $= m_{n+2}$. Thus $S(n)$ implies $S(n+1)$.

Section 8

1. By considering three cases: (i) $x \leq -1$, (ii) $-1 < x < 1$, (iii) $x \geq 1$, one finds that $|x + 1| < |x - 1|$ if and only if $x < 0$.

3. (a) The Division Theorem implies that for each integer a, there exist integers q and r such that $a = 3 \cdot q + r$ and $r = 0, 1$, or 2.
   (b) Suppose a and b are integers such that 3 divides $a \cdot b$ but 3 divides neither a nor b. Then there exist integers q and r such that $a = 3 \cdot q + r$ where $r = 1$ or 2 and integers p and s such that $b = 3 \cdot p + s$ where $s = 1$ or 2. Consider four cases: (1) $r = s = 1$; (2) $r = 1$, $s = 2$; (3) $r = 2$, $s = 1$; (4) $r = s = 2$.
   Case 1. If $r = s = 1$, then $a \cdot b = (3q+1) \cdot (3 \cdot p + 1) = 3(3 \cdot p \cdot q + p + q) + 1$. Since 3 divides both $a \cdot b$ and $3(3 \cdot p \cdot q + p + q)$, 3 divides 1, which is a contradiction. Thus Case 1 cannot hold. Similarly, the other cases can be eliminated. Thus if 3 divides $a \cdot b$, then 3 divides either a or b.

6. The bootstrap proof consists of a sequence of cases:

Case 1. $r = 0$: $\log(x^0) = \log(1) = 0 = 0 \cdot \log x$.

Case 2. r is a positive integer: The proof that $\log(x^r) = r \cdot \log(x)$ is by induction.

Case 3. r is a negative integer. Then $0 = \log(1) = \log(x^r \cdot x^{-r}) = \log(x^r) + \log(x^{-r}) = \log(x^r) + (-r) \cdot \log(x)$ by Case 2. Therefore, $\log(x^r) = r \cdot \log(x)$.

Case 4. $r = m/n$ where m and n are integers. Then $m\log(x) = \log(x^m) = \log\left((x^{m/n})^n\right) = n\log(x^{m/n}) = n\log(x^r)$. Therefore, $\log(x^r) = (m/n)\log(x) = r \cdot \log(x)$.

Section 9

2. $\sqrt{2}$ is irrational, yet $\sqrt{2} \cdot \sqrt{2}$ is rational.

4. $2 = \sqrt{1} + \sqrt{1} \neq \sqrt{2} = \sqrt{1+1}$. Thus the square root function does not have the additive property.

5. (a) If an even integer is multiplied by an odd integer, then the product is an even integer.

   (b) For all integers x and y, if x is even and y is odd, then $x \cdot y$ is even.

Section 10

3. (a) First, suppose $A \subseteq B$ and show $A \cup B = B$. To do so, we show (i) $A \cup B \subseteq B$ and (ii) $B \subseteq A \cup B$. (i) Let $x \in A \cup B$. We show that $x \in B$. Then $x \in A$ or $x \in B$; if $x \in A$, then, since $A \subseteq B$, $x \in B$; on the other hand, if $x \in B$, then $x \in B$. Thus, if $x \in A \cup B$, then $x \in B$. (ii) Let $x \in B$. Then certainly, $x \in A$ or $x \in B$. Thus $B \subseteq A \cup B$. Therefore, if $A \subseteq B$, then $A \cup B = B$.

   For the converse, suppose $A \cup B = B$ and show $A \subseteq B$. We must show that if $x \in A$, then $x \in B$. If $x \in A$, then $x \in A$ or $x \in B$, i.e., $x \in A \cup B$. But $A \cup B = B$ and hence $x \in B$. Therefore, if $A \cup B = B$, then $A \subseteq B$.

4. (a) $x \in A - B$ if and only if $x \in A$ and $x \notin B$ if and only if $x \in A$ and $x \in B^c$ if and only if $x \in A \cap B^c$. Thus $A - B = A \cap B^c = B^c \cap A = B^c \cap (A^c)^c = B^c - A^c$.

6. $P(\{1\}) = \{\emptyset, \{1\}\}$, $P(P(\{1\})) = \{\emptyset, \{\emptyset\}, \{\{1\}\}, \{\emptyset, \{1\}\}\}$.

8. (a) $C \in P(A \cap B)$ if and only if $C \subseteq A \cap B$ if and only if $C \subseteq A$ and $C \subseteq B$ if and only if $C \in P(A) \cap P(B)$.

   (b) $P(A) \cup P(B) \subseteq P(A \cup B)$ for all sets A and B. If $C \in P(A) \cup P(B)$, then $C \in P(A)$ or $C \in P(B)$. Thus $C \subseteq A$ or $C \subseteq B$, and, in either case, $C \subseteq A \cup B$, i.e., $C \in P(A \cup B)$.

13. (a) By definition, $(a, a) = \{\{a\}, \{a, a\}\}$. But for any x, $\{x, x\} = \{x\}$. Thus, $(a, b) = \{\{a\}, \{a\}\} = \{\{a\}\}$.

(b) $\{a\} \times \{a\} = \{(a, a)\} = \{\{\{a\}\}\}$ by part (a).

15. $A = \emptyset$ or $B = \emptyset$.

19. If $A = \emptyset$ and $B \neq C$, then $A \times B = A \times C$ and yet $B \neq C$. Salvage: If $A \neq \emptyset$, and $A \times B = A \times C$, then $B = C$. Proof: Since $A \neq \emptyset$, there exist $a \in A$. Then $y \in B$ if and only if $(a, y) \in A \times B = A \times C$ if and only if $y \in C$.

Section 11

1. $R_< \subset R_\leq, R_> \subset R_\geq, R_\leq \cap R_\geq = \{(x, x) \mid x \in \mathbf{R}\}, R_< \cap R_> = \emptyset, R_\leq \cap R_> = \emptyset$.

3. (a) There are four reflective relations on $\{1, 2\}$.

(b) There are eight symmetric relations on $\{1, 2\}$.

(c) There are two reflexive and symmetric relations on $\{1, 2\}$.

(d) There are six relations on $\{1, 2\}$ that are neither reflexive nor symmetric.

7. (a) $\emptyset$ is symmetric since the statement "if $(a, b) \in \emptyset$, then $(b, a) \in \emptyset$" is true, because its hypothesis is false. $\emptyset$ is transitive for the same reason.

9. (c) $R \cap S$ is symmetric: if $(a, b) \in R \cap S$, then $(a, b) \in R$ and $(a, b) \in S$. Since R and S are symmetric, $(b, a) \in R$ and $(b, a) \in S$ and hence $(b, a) \in R \cap S$.

Section 12

1. (c) $f : A \to B$ defined by $f(x) = b$ for all $x \in A$ is the only function from A to B.

(e) $f_1(1) = 1$, $f_2(1) = 2$, $f_3(1) = 3$ define the only functions from A to B.

2. (a) f is bijective.

(b) f is bijective.

(c)–(e) f is neither injective nor surjective.

4. (b) $\text{Ran}(f) = \mathbf{R}$.

(c) $f^{-1}(x) = (1/a)(x - b)$.

8. (a) There are two bijections from A to A.

10. (a) $f(f^{-1}(B_1)) \subseteq B_1$. Let $y \in f(f^{-1}(B_1))$. Then $y = f(x)$ for some $x \in f^{-1}(B_1)$. Thus $y = f(x) \in B_1$. Let $f : \mathbf{R} \to \mathbf{R}$ be given by $f(x) = 5$ for all $x \in \mathbf{R}$. Then $\{5\} = f(f^{-1}(\mathbf{R})) \subset \mathbf{R}$.

(b) $f^{-1}(f(A_1)) \supseteq A_1$. Let $x \in A_1$. Then $f(x) \in f(A_1)$, and hence $x \in f^{-1}(f(A_1))$. The example given in (a) shows that the inclusion can be proper.

11. (a) $(g \circ f)^{-1} = f^{-1} \circ g^{-1}$.

(d) We show that f is (i) injective and (ii) surjective. (i) Suppose $x_1, x_2 \in A$ with $f(x_1) = f(x_2)$. Then $(g \circ f)(x_1) = g(f(x_1)) = g((f(x_2)) = (g \circ f)(x_2)$. But $g \circ f$ is bijective and hence is injective. Thus $x_1 = x_2$, and f is injective. (ii) Let $y \in B$. Then we must find $x \in A$ such that $f(x) = y$. Let $z = g(y)$. Then $z \in C$; since $g \circ f : A \to C$ is bijective, there exists $x \in A$ such that $(g \circ f)(x) = z$. Since $g(y) = z = (g \circ f)(x) = g((f(x))$ and since g is injective, we conclude that $y = f(x)$.

15.(a) $f^{-1}(x) = f(x) = 1/x$.

Section 13

3. (a) Since for any set A, I_A and $A \times A$ are equivalence relations on A, these equivalence relations can be labeled as "trivial."

(b) If A has only one equivalence relation, then $I_A = A \times A$ and hence A contains exactly one element.

5. The relations defined in (a) and (c) are equivalence relations. The others are not.

7. $\sim$ is an equivalence relation on $C[a, b]$.

10. (b) $[1/2]_{\equiv_1} = \{x \in \mathbf{R} \mid x = n + 1/2 \text{ for some } n \in \mathbf{Z}\}$.

13. (b) $[(16, 1)] = \{(16, 1), (4, 2), (2, 4)\}$
$[(3, 4)] = \{(3, 4), (81, 1)(9, 2)$
(c) $[(64, 1)] = \{(64, 1), (8, 2), (4, 3), (2, 6)\}$

Section 15

1. The unknown is a pair of numbers a, b such that $a > 0, b > 0, a+b = 10$, and $a \cdot b$ is as large as possible. Another description: Since $a + b = 10$, $b = 10 - a$. Thus the unknown is a number a such that $0 < a < 10$ and $a \cdot (10 - a)$ is as large as possible.

4. The unknown consists of the collection of all vectors $\mathbf{v}$ such that $\mathbf{v}$ is perpendicular to $1, 2, 3$. In algebraic terms, the unknown is the set of all vectors $\mathbf{v} = (x, y, z)$ such that

$$\mathbf{v} \cdot (1, 2, 3) = x + 2y + 3z = 0.$$

6. The unknown is $\int_0^\pi \sin(x)\, dx$.

9. (a) Algebraic representation: Find (x, y, z) such that $x - 3y - z = 0 = y + 2z$.

   (b) Find all (x, y, z, w) such that $x + 2y - z + 2w = 0 = y + 2z + 3w$.

Section 16

2. Related problem: Find a nonzero vector $\mathbf{v}$ such that $\mathbf{v}$ makes an angle of $\pi/6$ with $(2, 3)$. Another: Find a unit vector $\mathbf{u}$ that makes an angle of $\pi/2$ with $(2, 3)$.

3. Related problem: Prove that the infinite series $\sum_{n=1}^{\infty} 1/n^2$ converges.

6. (a) Related problem: Evaluate $\sum_{1}^{\infty} \left(\dfrac{1}{n} - \dfrac{1}{n+1}\right)$.

   (b) Related problem: Evaluate $\sum_{n=1}^{k} \dfrac{n}{(n+1)!}$ for each positive integer k.

10. Generalization: Suppose n is a positive integer and that $3 \cdot n$ points in the plane are given such that no three of the points are collinear. Prove that there exist n disjoint triangles having the given points vertices.

11. Any set of $n + 1$ vectors in $\mathbf{R}^n$ is linearly dependent.

References

1. P. Davis and R. Hersh, *The Mathematical Experience*, Birkhauser, 1982.

2. S. Galovich, *Introduction to Mathematical Structures*, Harcourt Brace Jovanovich, 1989.

3. A. G. Hamilton, *Logic for Mathematicians*, Cambridge University Press, 1978.

4. A. G. Hamilton, *Numbers, Sets and Axioms*, Cambridge University Press, 1982.

5. J. Mason, L. Burton, and K. Stacey, *Thinking Mathematically*, Addison-Wesley, 1985.

6. G. Polya, *How to Solve It*, Princeton University Press, 1945.

7. G. Polya, *Mathematical Discovery* (2 vols.), John Wiley, 1962 (Vol. I), 1965 (Vol. II).

8. G. Polya, *Mathematics and Plausible Reasoning* (2 vols.), Princeton University Press, 1954 (Vol. I), 1968 (Vol. II).

9. A. Schoenfeld, *Mathematical Problem Solving*, Academic Press, 1985.

Index

A

analogy 161–162
associative law 51, 82
axiom system 48–50

B

biconditional 2, 8
binary string 18
bootstrap argument 71
boundedness 52
broadly modified prob-
lems 160 *ff*

C

Cartesian product 91 *ff*
case analysis 71 *ff*
commutative law 51, 82
complement 20
completeness 52
composition of functions 114
conditional 2, 6
congruence modulo n 130
conjunction 2, 3
containment of sets 15
contradiction 10, 43
contrapositive 7
converse 8
counterexamples 77 *ff*

D

DeMorgan's laws 86
disjunction 2, 3
distributive law 40, 51, 82
divide-and-conquer 71
division algorithm 66
divisor 41
domain of a function 108

E

element of a set 13

empty (null) set 14
equality of sets 14
equivalence class 132
equivalence relation 102, 126 *ff*
essentially equivalent problems 155 *ff*
examine special cases 149 *ff*
existential quantifier 28

F

factor 40
Fibonacci sequence 109
field
axioms 50–51
of real numbers 50 *ff*
ordered 57
finite set 14
focusing on the unknown 143 *ff*
function 102, 103, 108
bijective or one-to-one
correspondence 110
injective or one-to-one 110
surjective or onto 110

G

generalization 163–165

H

heuristics 140 *ff*

I

idempotent law 82
implication 6
infinite set 14
integer 23
intersection of sets 19
inverse function 116
inverse image 120
irrational number 24, 43
isolate the hypothesis 145

isomorphic   53

L

least upper bound   52
logical equivalence   5
looking back   166–168
Lucas sequence   121, 180

M

mathematical induction   61 *ff*
model   53

N

natural number   22
negation   2, 4
null (empty) set   14

O

operations on sets
  complement   20
  difference   19
  intersection   19
  union   19
ordered field   56
ordered pair   91
ordering   52

P

partial ordering   101
partition   134
postulates   48
power set   87
predicate   27
prime integer   44
problem representation   141 *ff*
proof methods
  case analysis   71 *ff*
  contradiction   43 *ff*
  contraposition   42
  direct   38 *ff*
  mathematical induction   61 *ff*
proposition   2

R

range of a function   108
rational number   23
real number   22 *ff*, 50 *ff*
recursive sequence   123
related problems   155 *ff*
relation   99
  antisymmetric   100
  equivalence   102
  identity   99
  reflexive   100
  symmetric   100
  transitive   100

S

set
  complement   20
  difference   19
  finite   14
  infinite   14
set modulo an equivalence relation   134
simplify the problem   151 *ff*
singleton   14
slightly modified prob-
  lems   159 *ff*
strong induction   67
subset   15

T

tautology   10

U

union of sets   19
universal quantifier   28
upper bound   52

V

variable   27
Venn diagram   17